Lecture Notes in Computer Science 13050

Tanveer Syeda-Mahmood · Xiang Li ·
Anant Madabhushi · Hayit Greenspan ·
Quanzheng Li · Richard Leahy · Bin Dong ·
Hongzhi Wang (Eds.)

Multimodal Learning for Clinical Decision Support

11th International Workshop, ML-CDS 2021
Held in Conjunction with MICCAI 2021
Strasbourg, France, October 1, 2021
Proceedings

Editors
Tanveer Syeda-Mahmood (iD)
IBM Research - Almaden
San Jose, CA, USA

Anant Madabhushi (iD)
Case Western Reserve University
Cleveland, OH, USA

Quanzheng Li (iD)
Harvard Medical School
Boston, MA, USA

Bin Dong
Peking University
Beijing, China

Xiang Li (iD)
Harvard Medical School
Boston, MA, USA

Hayit Greenspan (iD)
Tel Aviv University
Tel Aviv, Israel

Richard Leahy
University of Southern California
Los Angeles, CA, USA

Hongzhi Wang (iD)
IBM Research - Almaden
San Jose, CA, USA

ISSN 0302-9743   ISSN 1611-3349 (electronic)
Lecture Notes in Computer Science
ISBN 978-3-030-89846-5   ISBN 978-3-030-89847-2 (eBook)
https://doi.org/10.1007/978-3-030-89847-2

LNCS Sublibrary: SL6 – Image Processing, Computer Vision, Pattern Recognition, and Graphics

This Springer imprint is published by the registered company Springer Nature Switzerland AG
The registered company address is: Gewerbestrasse 11, 6330 Cham, Switzerland

Preface

On behalf of the organizing committee, we welcome you to the proceedings of the Eleventh Workshop on Multimodal Learning and Fusion Across Scales for Clinical Decision Support (ML-CDS 2021), which was held virtually at MICCAI 2021 in Strasbourg, France. This year's edition reflects a confluence of two workshops, namely, Multimodal Learning for Clinical Decision Support, which has been running for the last 10 years at MICCAI, and Multiscale Multimodal Medical Imaging, which was held at MICCAI 2019. Overall, the goal of this series of workshops has been to bring together medical image analysis and machine learning researchers with clinicians to tackle the important challenges of acquiring and interpreting multimodality medical data at multiple scales for clinical decision support and treatment planning, and to present and discuss latest developments in the field.

The previous workshops on this topic have been well-received at MICCAI, specifically Lima (2020), Shenzen (2019), Granada (2018), Quebec City (2017), Athens (2016), Munich (2015), Nagoya (2013), Nice (2012), Toronto (2011), and London (2009). Continuing on the momentum built by these workshops, this year's edition focused on integrating diagnostic imaging, pathology imaging, and genomic datasets for diagnosis and treatment planning, treating clinical decision support on a holistic basis.

The workshop received a total of 16 submissions. All submissions underwent a double-blind peer-review process, with each submission being reviewed by at least two independent reviewers and one Program Committee member. Based on the review scores and comments, 10 papers were accepted for presentation at the workshop, which are included in this Springer LNCS volume. We would like to thank the authors for their submissions and all the Program Committee members for handling the submissions with professional judgement and constructive comments.

With less than 5% of medical image analysis techniques translating to clinical practice, workshops on this topic have helped raise the awareness of our field to clinical practitioners. The approach taken in the workshop is to scale it to large collections of patient data exposing interesting issues of multimodal learning and its specific use in clinical decision support by practicing physicians. The ultimate impact of these methods can be judged when they begin to affect treatment planning in clinical practice.

We hope that you enjoyed the program we assembled and, for those readers who were able to participate in the workshop, the discussion on the topics of the papers and the panel.

October 2021

Tanveer Syeda-Mahmood
Xiang Li

Organization

General Chairs

Tanveer Syeda-Mahmood   IBM Research Almaden, USA
Xiang Li   Massachusetts General Hospital, USA

Program Chairs

Anant Madabhushi   Case Western Reserve University, USA
Hayit Greenspan   Tel-Aviv University, Israel
Quanzheng Li   Massachusetts General Hospital, USA
Richard M. Leahy   University of Southern California, USA
Bin Dong   Peking University, China
Hongzhi Wang   IBM Research Almaden, USA

Program Committee

Amir Amini   University of Louisville, USA
Sameer Antani   National Library of Medicine, USA
Rivka Colen   MD Andersen Cancer Center, USA
Keyvan Farahani   National Cancer Institute, USA
Alejandro Frangi   University of Sheffield, UK
Guido Gerig   University of Utah, USA
David Gutman   Emory University, USA
Allan Halpern   Memorial Sloan Kettering Research Center, USA
Ghassan Hamarneh   Simon Fraser University, Canada
Jayshree Kalpathy-Kramer   Massachusetts General Hospital, USA
Ron Kikinis   Harvard University, USA
Georg Langs   Medical University of Vienna, Austria
B. Manjunath   University of California, Santa Barbara, USA
Dimitris Metaxas   Rutgers University, USA
Nikos Paragios   CentraleSupélec, France
Daniel Racoceanu   National University of Singapore, Singapore
Eduardo Romero   Universidad Nacional de Colombia, Colombia
Daniel Rubin   Stanford University, USA
Russ Taylor   Johns Hopkins University, USA
Agma Traina   University of São Paulo, Brazil
Max Viergever   Utrecht University, The Netherlands
Sean Zhou   Siemens Corporate Research, USA

Contents

Merging and Annotating Teeth and Roots from Automated Segmentation
of Multimodal Images   81
  Romain Deleat-Besson, Celia Le, Winston Zhang, Najla Al Turkestani,
  Lucia Cevidanes, Jonas Bianchi, Antonio Ruellas, Marcela Gurgel,
  Camila Massaro, Aron Aliaga Del Castillo, Marcos Ioshida,
  Marilia Yatabe, Erika Benavides, Hector Rios, Fabiana Soki,
  Gisele Neiva, Kayvan Najarian, Jonathan Gryak, Martin Styner,
  Juan Fernando Aristizabal, Diego Rey, Maria Antonia Alvarez,
  Loris Bert, Reza Soroushmehr, and Juan Prieto

Structure and Feature Based Graph U-Net for Early Alzheimer's Disease
Prediction   93
  Yun Zhu, Xuegang Song, Yali Qiu, Chen Zhao, and Baiying Lei

A Method for Predicting Alzheimer's Disease Based on the Fusion
of Single Nucleotide Polymorphisms and Magnetic Resonance Feature
Extraction   105
  Yafeng Li, Yiyao Liu, Tianfu Wang, and Baiying Lei

Author Index   117

From Picoscale Pathology to Decascale Disease: Image Registration with a Scattering Transform and Varifolds for Manipulating Multiscale Data

Kaitlin M. Stouffer[1]([✉])(iD), Zhenzhen Wang[1], Eileen Xu[1], Karl Lee[1], Paige Lee[2], Michael I. Miller[1], and Daniel J. Tward[2](iD)

[1] Johns Hopkins University, Baltimore, MD 21218, USA
{kstouff4,zwang218,exu1,slee508,mim}@jhu.edu
[2] University of California Los Angeles, Los Angeles, CA 90095, USA
paigetlee1@g.ucla.edu, dtward@mednet.ucla.edu

Abstract. Advances in neuroimaging have yielded extensive variety in the scale and type of data available. Effective integration of such data promises deeper understanding of anatomy and disease–with consequences for both diagnosis and treatment. Often catered to particular datatypes or scales, current computational tools and mathematical frameworks remain inadequate for simultaneously registering these multiple modes of "images" and statistically analyzing the ensuing menagerie of data. Here, we present (1) a registration algorithm using a "scattering transform" to align high and low resolution images and (2) a varifold-based modeling framework to compute 3D spatial statistics of multiscale data. We use our methods to quantify microscopic tau pathology across macroscopic 3D regions of the medial temporal lobe to address a major challenge in the diagnosis of Alzheimer's Disease–the reliance on invasive methods to detect microscopic pathology.

Keywords: Scattering transform · Varifold · Alzheimer's disease

1 Introduction

Current imaging technologies capture data of diverse types at varying spatial/temporal resolutions. Often combined and clinically available, these imaging modalities yield multi-faceted, high dimensional datasets that hold promise for diagnosis and management of neurological disease [15]. However, non-invasive imaging methods only capture patterns at macroscopic to mesoscopic levels. Microscopic markers, as identified through immunohistochemistry and other histological staining procedures, though crucial to the understanding and identification of diseases, have not been adequately linked to the measures of these other non-invasive imaging types. Recent strides to digitize pathological samples [21] and "image" microscopic data, such as RNA [28], herald the opportunity to

© Springer Nature Switzerland AG 2021
T. Syeda-Mahmood et al. (Eds.): ML-CDS 2021, LNCS 13050, pp. 1–11, 2021.
https://doi.org/10.1007/978-3-030-89847-2_1

develop image registration and analysis tools that handle images across the *full* spectrum of microscopic to macroscopic scales. The linking of this data will not only provide a more coherent, 3D picture of anatomy from cells to tissues, but could establish a cellular basis for many macroscopic biomarkers as signatures of type and stage of disease.

We focus our method development on linking MRI biomarkers for Alzheimer's Disease (AD) to its well-established tau signatures [11]. AD continues to be diagnosed from tau and β-amyloid ($A\beta$) patterns [4] in histological tissue samples, recruited often postmortem. Recently, thickness changes in transentorhinal cortex (TEC) have been identified that posit tremendous potential as biomarkers, particularly for early, pre-clinical periods of AD. They vary with stage of cognitive decline and are identifiable through non-invasive means (MRI) [13]. We hypothesize that these thickness changes reflect underlying changes in tau and $A\beta$ pathology, and ultimately aim to show this through successful registration and joint statistical analysis of digital pathology and MR images of tissue from the medial temporal lobe (MTL) (see Fig. 1A,B).

Amongst its challenges, analysis of data at multiple scales and of multiple modes requires (1) alignment of this data into a common physical coordinate space and (2) a mathematical framework for representing and manipulating diverse datatypes at different scales. To address (1), we situate our imaging data in the coordinate space of the Mai-Paxinos atlas, well-known globally to neuropathologists [16]. Additionally, we use a "Scattering Transform" [18] in our registration approach to capture textural information of high resolution images and thereby facilitate registration between images of different modalities and scales. To address (2), we model our data with varifolds [1]: sets of discrete particles with associated weights, physical locations, and feature distributions of arbitrary dimension, which accommodate diverse image types. By defining a set of particles per image, sampled in space according to the image's resolution and with weights reflective of initial mass or "observation area", we accommodate regions of sparse vs. dense observations in the unified Mai atlas space–a distinction modern methods may fail to capture. Furthermore, we can compute 1st, 2nd, and higher order moments of any of the features at any scale.

In the following sections, we detail these described methods and demonstrate their effectiveness at combining and consolidating microscopic histolopathological data with macroscopic MRI data from a case of advanced AD.

2   Methods

2.1   Image Acquisition and Processing

Imaging data was compiled by the Johns Hopkins University Brain Resource Center and the laboratory of Dr. Susumu Mori. Samples were prepared from a 1290 g brain of a 93-year-old male with a clinical diagnosis of AD and extensive neurofibrillary pathology, characterized as Braak stage VI [4] in autopsy. Three contiguous blocks of tissue, sized 20–30 mm in height, width, and depth were excised from the MTL. Blocks were imaged individually with an 11.7T MRI scanner at 0.125 mm isotropic resolution and then sectioned into 15 slices, which were 10 microns thick, spaced 1 mm apart, and stained for phosphorylated tau (PHF-1). Stained slices were digitized at 2 micron resolution.

Machine learning, and particularly neural networks, have become common-place in detection algorithms for AD pathology [12,25]. We detected individual tau tangles in histology images (see Fig. 1A) using a multi-step convolutional neural network-based procedure. We trained a UNET [23] with the architec-ture outlined in Table 1 on a dataset comprised of subsets of pixels manually annotated with a binary label (tau/no tau) according to staining intensity and corroborated by a neuropathologist. We used our trained UNET to generate per pixel probabilities of tau on all histology slices. Finally, we used an available marker-based implementation of the "watershed algorithm" [5] to detect indi-vidual tau tangles from these probabilities, as connected high probability units. Similar approaches have reported accuracy above 0.95 [26,29]. Centers of each tau tangle were added as features to histology images.

Under the guidance of a neuropathologist, we manually segmented aligned blocks of 3D MRI into subregions of MTL including amygdala, entorhinal cortex (ERC), cornu ammonis fields (CA1, CA2, CA3), and subiculum using Seg3D [8]. Per voxel segmentation labels were added as features to MRIs.

Table 1. Structure of UNET trained to detect tau tangles. Contraction layers are shown in the left 3 columns, and expansion layers in the right 3 columns. Number of parameters listed correspond to linear filters + bias vector. Conv: 3×3 Convolution with stride 1, MP: 2×2 max pool, ReLU: Rectified Linear Unit , ConvT: 2×2 transposed convolution with stride 2. Note number of features doubles in the expansion layers due to concatenation with the contraction layers (skip connections).

No.	Contract	Parameters	No.	Expand	Parameters
1	Conv	$8 \times 3 \times 3 \times 3 + 8$	18	ConvT	$32 \times 64 \times 2 \times 2 + 32$
2	Conv	$8 \times 8 \times 3 \times 3 + 8$	19	Conv	$32 \times 64 \times 3 \times 3 + 32$
3	MP	0	20	ReLU	0
4	Conv	$16 \times 8 \times 3 \times 3 + 16$	21	Conv	$32 \times 32 \times 3 \times 3 + 32$
5	ReLU	0	22	ReLU	0
6	Conv	$16 \times 16 \times 3 \times 3 + 16$	23	ConvT	$16 \times 32 \times 2 \times 2 + 16$
7	ReLU	0	24	Conv	$16 \times 32 \times 3 \times 3 + 16$
8	MP	0	25	ReLU	0
9	Conv	$32 \times 16 \times 3 \times 3 + 32$	26	Conv	$16 \times 16 \times 3 \times 3 + 16$
10	ReLU	0	27	ReLU	0
11	Conv	$32 \times 32 \times 3 \times 3 + 32$	28	ConvT	$8 \times 16 \times 2 \times 2 + 8$
12	ReLU	0	29	Conv	$8 \times 16 \times 3 \times 3 + 8$
13	MP	0	30	ReLU	0
14	Conv	$64 \times 32 \times 3 \times 3 + 64$	31	Conv	$8 \times 8 \times 3 \times 3 + 8$
15	ReLU	0	32	ReLU	0
16	Conv	$64 \times 64 \times 3 \times 3 + 64$	33	Linear	$2 \times 8 + 2$
17	ReLU	0			
				Total	120,834

Fig. 1. (A) PHF-1 histology slice with detected tau tangles (red circles) in $1\,\mathrm{mm}^2$. (B) 2D image slice of 3D MRI with manual segmentations of MTL regions. (C) Scattering Transform of histology image showing cellular texture. Subspace spanned by principle components corresponding to 3 largest eigenvalues mapped to RGB. (Color figure online)

2.2   Algorithm for Multimodal Registration with Damaged Tissue

Many methods have been developed to align 2D to 3D images. They vary in such choices as transformation model (rigid vs. non-rigid) and optimization procedure (continuous vs. discrete) [10,22]. To transform histological and MRI data into the coordinate space of the Mai-Paxinos atlas, we build on an approach used for registering similar images of mouse brain tissue [27]. In this approach, the authors estimate a series of 3D and 2D geometric transformations together with a contrast transform to minimize a cost computed between single blocks of MRI and the *set* of corresponding histology slices.

Here, we modify this approach by separating 2D and 3D transformation parameter estimation to improve 2D estimation with a "Scattering Transform" [18] (see Sect. 2.3). 3D parameters are estimated as in [27] and include an affine transformation (A_2) with 8 degrees of freedom (including rigid, slice thickness, and pixel size parameters) and diffeomorphism (φ_2)–computed from a smooth time-varying velocity field using large deformation diffeomorphic metric matching (LDDMM) [2]. 2D parameters are estimated by minimizing, subject to a regularization term, the *weighted* sum of square error between deformed Scatter Transform images (downsampled histology with 6 channels) and estimated corresponding 2D slices of MRI. The parameters include a diffeomorphism (φ_1), rigid transformation with scaling (A_1) with 4 degrees of freedom, and contrast transform (F, 1st degree polynomial) with 7 degrees of freedom. Weights equal the posterior probability of MRI pixels pertaining to "matching" tissue, "background" (dark), or "artifacts" (bright). They are computed with Bayes Theorem assuming a Gaussian distribution for each category with fixed variance and means: $\sigma_M = 0.2, \sigma_B = 0.5, \sigma_A = 5\sigma_M$, $\mu_A = 2.0, \mu_B = 0.0$ (intensity units), and μ_M estimated jointly with deformation parameters using the Expectation-Maximization (EM) algorithm [9].

Remaining transformations (see Table 2) include manually determined rigid alignments between single block MRIs (A_3) and between surface representations of MRI segmentations and the atlas (A_4).

Table 2. Transformations estimated in registration of 2D histology to the Mai atlas.

Transformation	Domain and Range Space	Type
$\varphi_1 : \mathbf{x} \mapsto \varphi_1(\mathbf{x})$	$\mathbb{R}^2 \to \mathbb{R}^2$ (Downsampled Histology, DH)	*diffeomorphism*
$A_1 \in \mathbb{R}^{3x3}$	$\mathbb{R}^2 \to \mathbb{R}^2$ (DH)	*rigid+scale*
$S : (x, y) \mapsto (x, y, z), z \in Z$	$\mathbb{R}^2 \to \mathbb{R}^3$ (DH to Block MRI)	*slice stacking*
$A_2 \in \mathbb{R}^{4x4}$	$\mathbb{R}^3 \to \mathbb{R}^3$ (Block MRI)	*rigid+(xy)/z scale*
$\varphi_2 : \mathbf{x} \mapsto \varphi_2(\mathbf{x})$	$\mathbb{R}^3 \to \mathbb{R}^3$ (Block MRI)	*diffeomorphism*
$A_3 \in \mathbb{R}^{4x4}$	$\mathbb{R}^3 \to \mathbb{R}^3$ (Block MRI to Combined MRI)	*rigid*
$A_4 \in \mathbb{R}^{4x4}$	$\mathbb{R}^3 \to \mathbb{R}^3$ (Combined MRI to Mai)	*rigid*
$F : \mathbf{I} \mapsto F(\mathbf{I})$	$\mathbb{R}^6 \to \mathbb{R}$ (Scattering to Grayscale)	*polynomial*

2.3   Scattering Transform for Retaining High Resolution Texture

The Scattering Transform was first introduced by Mallat [18] as a means of capturing frequency information of signals (images) in a manner Lipschitz continuous to small diffeomorphisms. Its structure is equivalent to some forms of classical convolutional neural networks [18], and it has been harnessed in a range of image processing and classification tasks such as digit recognition and texture discrimination [6], without need for training. Here, we compute the discrete *subsampled* windowed Scattering Transform of high resolution images to downsample histology images whilst maintaining key discriminatory textural information. This textural information is analogous to features extracted in current practices of radiomics [14]. We follow an approach similar to the "Filterbank Algorithm" as in [17,24].

The transform consists of a sequence (path) of wavelet convolutions at increasing scales, each followed by a simple nonlinearity (modulus) and downsampling operation. For image registration, we require orientation invariance, and so, in our approach, we use highpass filters instead of wavelets (identity minus a convolution with a Gaussian of width 5 pixels). Other strategies of achieving orientation and general rigid-motion invariance are discussed in [24]. As in [18], we consider paths that have at most 2 nonlinearities–sufficient for capturing the spectrum of texture in most "natural images"– and downsample 5 times by a factor of 2 to reach the resolution of MRI. The resulting downsampled histology images have 48 channels, each corresponding to the Scattering Transform coefficient of a distinct path. Since channels are highly correlated (i.e. all respond strongly to edges), we use PCA to choose a 6 dimensional subspace (see Fig. 1C), which reduces computational complexity and mitigates overfitting.

2.4   Varifold Measures for Modeling and Crossing Multiple Scales

To quantify features of our deformed imaging data, we model our data with varifolds. First described by Almgren and later Allard [1], varifolds were introduced into Computational Anatomy to describe shapes (surfaces) as distributions of unoriented tangent spaces [7]. The formulation lends itself to capturing not just physical orientation but orientation in any (feature) space and models data in continuous and discrete settings equally well with a Dirac delta function. As such, varifolds can model smooth functional data observed at tissue scales (i.e. cortical thickness) and discrete particle data observed at cellular scales (i.e. tau tangle locations).

Following recent work extending the use of varifolds in Computational Anatomy [19,20], we model our multimodal, multiscale imaging data with a hierarchy of discrete varifolds across scales (resolutions). At the finest scale, we define particles as pixels of histology images. Similar to the approach described in [7], we treat each histology image as a 2D surface in 3D where pixels are quadrilateral faces. Weights associated with each particle are equal to the area of each face. Features include MTL subregion and presence of tau (see Table 3). Particle locations are carried to Mai space under the action of our composition of 2D and 3D transformations (see Table 2), and weights change according to the 2D Jacobian of this composition altering the 2D subspace spanned by each face (see Fig. 2). The resulting weights continue to reflect 2D histology tissue area.

Table 3. Fine particle features for PHF-1 stained histology slices. L denotes total regions (12 MTL subregions and other).

Feature	Description
$\tau \in \mathbb{R}$	Number of tau tangles per square mm of tissue
$\mathbf{t} \in \mathbb{R}^L$	Number of tau tangles in each region per square mm of tissue
$\mathbf{r} \in [0,1]^L$	Square mm of each region per square mm of tissue

To quantify statistical trends of features over 3D space at the resolution of MRI, we resample the Mai space at a coarser lattice of particles and define spatial and feature kernels that govern the contribution of fine particles to coarse ones. Here, we resample at an intermediate level without smoothing (i.e. rectangular kernel) and at the coarsest level with either an isotropic 3D Gaussian of variance 1 mm (see Fig. 2) or skewed 3D Gaussian (see Fig. 4) in the respective cases of coarse particles sampled equally along each dimension or along a single axis. We combine fine particle feature contributions with 1st and 2nd central moments (see Fig. 2D,E).

3   Results

We use our described methods to achieve two goals. First, we register 2 micron resolution 2D histology images of MTL tissue with corresponding 0.125 mm resolution 3D MRI. Figure 3 highlights our 2D transformation estimation, introduced here with use of the Scattering Transform to better predict white/gray matter contrast. We evaluate accuracy of our alignment by visual inspection of image overlap (Fig. 3A-D) and similar to others [29], by overlap in MRI segmentations

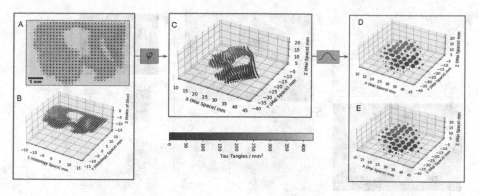

Fig. 2. Varifold particle representation of histology data. Circle area proportional to particle weight (2D tissue area). Circle color denotes feature (tau tangles/mm²). (A,B) 2D and 3D plots of fine particles in histology space. (C) Fine particles with transformed locations and weights. (D,E) Coarse particle resampling in Mai space with isotropic 3D Gaussian. 1st (D) and 2nd (E) central moments of tau tangles/mm².

deformed from 3D to 2D vs. original segmentations manually labeled in 2D histology slices (Fig. 3E-G). The success of remaining elements of our registration algorithm are demonstrated in Fig. 4B,C (i.e. transformation to Mai-Paxinos atlas space).

Second, we compute quantitative measures of tau tangle density. Figures 4D,E exhibit the outcome of our use of varifolds to sample and quantify tau tangle information along the anterior-posterior axis of the human brain (Mai z-axis). We use the weights in the varifold formulation to reflect how much histological tissue is observed at each coordinate. Circle area or bar transparency in Figs. 4D and 4E respectively indicate regions of more tissue area (larger circle, more opaque) vs. regions of less tissue area sampled (smaller circle, more transparent). We also harness the range of manipulations across diverse feature types available to us through kernel design by considering overall tau density vs. region specific tau density in Figs. 4D and 4E respectively.

4 Discussion

We have developed (1) a registration algorithm for aligning images of different modalities and (2) a mathematical architecture for representing and summarizing data of different types and spatial scales. As one application of our methods to the field of clinical neuroscience, we have demonstrated success at (1) registering 2D histology images with 3D MRI and the Mai-Paxinos atlas and (2) quantifying tau tangles, identified at the histological scale, with density measures along the anterior-posterior axis of the MTL, using the coordinates of the Mai atlas. This particular trend has been difficult to glean from analysis of histology alone.

Compared to other approaches for quantifying histological information in 3D, ours harbors a number of strengths. Our use of the Scattering Transform parallels machine learning tactics others have used to capture "featural" information

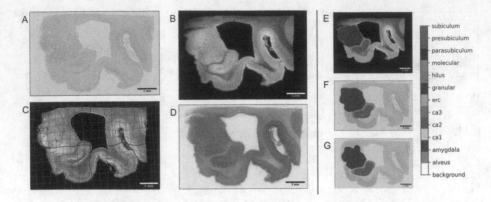

Fig. 3. 2D registration results using the Scattering Transform. (A) Original tau-stained histology image. (B) Estimated corresponding 2D MRI slice. (C) Deformed histology image (A) to MRI space via geometric transformations (φ_1, A_1) (gridlines) and contrast transform (F). (D) Deformed 2D MRI slice $(A_1^{-1}, \varphi_1^{-1})$, contrast inverted and overlaid over (A). (E) 2D MRI slice with MRI segmentations. (F) Deformed MRI segmentations overlaid on histology. (G) Manual histology segmentations for comparison to (F).

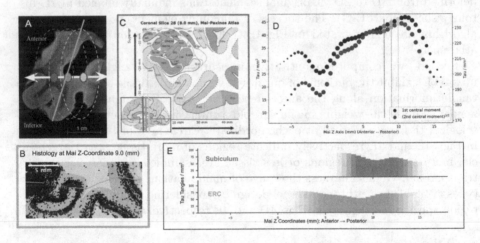

Fig. 4. (A) Sagittal slice at rostral end of 3D MRI with estimated axis and position of histology in (B). (B) Centers of tau tangles (black) and contour for Mai z-axis coordinate post deformation of histology. (C) Mai atlas, coronal slice 28 at $z = 8.0$ mm [16]. Tau tangle density overall (D) and within regions implicated in AD (E) computed at 0.5 mm intervals along Mai z-axis (anterior-posterior) from rostral MRI block. Area of circles (D) and transparency of bars (E) proportional to 2D histological tissue area.

of high resolution histology images [29], yet is more efficiently computed and generalizes to additional datasets without need of retraining. With our hierarchy of varifolds, we can carry and compute *quantitative* rather than categorical, relative, or semi-quantitative measures of pathology into 3D space, as published previously [3,29]. We can also discriminate areas of "missing tau" from those of missing data (i.e. tissue between histology sections) through the assignment of weights to our data, reflective of the amount of tissue area sampled. While others estimate segmentations from atlases [29], we have manually segmented MTL regions directly on our MRIs/histology images, which allows us to evaluate our algorithm *and* quantify tau in those regions more accurately. Finally, by deforming all image data to the space of the Mai-Paxinos atlas, we enhance interpretability of our results for a wider spectrum of biologists and neurologists.

Limitations of the work presented here include aspects of methods and data. First, our 2D transformation estimation assumes complete data for histology with artifacts and missing tissue defined only for MRI. Second, rigid alignments between MRI and the Mai-Paxinos atlas suffer from variation, particularly in collateral sulcus type, of our brain samples and the one in the atlas. To address these, we are developing an alternative registration method to allow for defects in all images and estimate nonrigid transformations between MRI and the atlas. Third, measures of tau tangle "density" are described in units of particles per square mm of tissue, and thus, are comparable only across histological samples of similar slice thickness (10 microns). Future work in this area includes the development of finer measures (i.e. "density" per unit volume) and discrimination between types of pathological tau (i.e. neurites vs. tangles). Finally, the results presented here reflect analysis of a single brain sample with advanced AD. We are currently using our methods to analyze brains with varying stages of AD and to compute both tau tangle and Aβ plaque density. This analysis will yield better assessment of the accuracy of our methods while bringing us closer to our clinical goal of linking MRI biomarkers [13] to underlying pathology.

Acknowledgements. This work was supported by the NIH (T32 GM 136577 (KS), U19 AG 033655 (MM), and R01 EB 020062 (MM)), the Kavli Neuroscience Discovery Institute (MM, DT), and the Karen Toffler Charitable Trust (DT). MM owns a founder share of Anatomy Works with the arrangement being managed by Johns Hopkins University in accordance with its conflict of interest policies. The remaining authors declare that the research was conducted in the absence of any commercial or financial relationships that could be construed as a potential conflict of interest. We thank Juan Troncoso, Susumu Mori, and Atsushi Saito at the Johns Hopkins Brain Resource Center for preparation of tissue samples. We thank Menno Witter at the Kavli Institute for Systems Neuroscience and Norwegian University of Science and Technology for his input with MTL segmentations.

References

1. Allard, W.K.: On the first variation of a varifold. Ann. Math. **95**(3), 417–491 (1972). https://doi.org/10.2307/1970934
2. Beg, M.F., Miller, M.I., Trouvé, A., Younes, L.: Computing large deformation metric mappings via geodesic flows of diffeomorphisms. Int. J. Comput. Vis. **61**(2), 139–157 (2005). https://doi.org/10.1023/B:VISI.0000043755.93987.aa
3. Blanken, A.E., et al.: Associations between hippocampal morphometry and neuropathologic markers of Alzheimer's disease using 7 T MRI. NeuroImage Clin. **15**, 56–61 (2017). https://doi.org/10.1016/j.nicl.2017.04.020

4. Braak, H., Braak, E.: Neuropathological stageing of Alzheimer-related changes. Acta Neuropathol. **82**(4), 239–259 (1991). https://doi.org/10.1007/BF00308809
5. Bradski, G.: The OpenCV Library. Dr. Dobb's J. Softw. Tools (2000)
6. Bruna, J., Mallat, S.: Invariant scattering convolution networks. IEEE Trans. Pattern Anal. Mach. Intell. **35**(8), 1872–1886 (2013). https://doi.org/10.1109/TPAMI.2012.230
7. Charon, N., Trouvé, A.: The varifold representation of nonoriented shapes for diffeomorphic registration. SIAM J. Imaging Sci. **6**(4), 2547–2580 (2013). https://doi.org/10.1137/130918885
8. CIBC seg3D: Volumetric image segmentation and visualization. scientific computing and imaging institute (SCI) (2016). http://www.seg3d.org
9. Dempster, A.P., Laird, N.M., Rubin, D.B.: Maximum likelihood from incomplete data via the EM algorithm. J. R. Stat. Soc. Series B **39**(1), 1–38 (1977)
10. Ferrante, E., Paragios, N.: Slice-to-volume medical image registration: a survey. Med. Image Anal. **39**, 101–123 (2017). https://doi.org/10.1016/j.media.2017.04.010
11. Jack, C.R., et al.: NIA-AA research framework: toward a biological definition of Alzheimer's disease. Alzheimer's Dement. **14**(4), 535–562 (2018). https://doi.org/10.1016/j.jalz.2018.02.018
12. Komura, D., Ishikawa, S.: Machine learning methods for histopathological image analysis. Comput. Struct. Biotechnol. J. **16**, 34–42 (2018). https://doi.org/10.1016/j.csbj.2018.01.001
13. Kulason, S., et al.: Cortical thickness atrophy in the transentorhinal cortex in mild cognitive impairment. NeuroImage Clin. **21**, 101617 (2019). https://doi.org/10.1016/j.nicl.2018.101617
14. Lambin, P., et al.: Radiomics: the bridge between medical imaging and personalized medicine. Nat. Rev. Clin. Oncol. **14**(12), 749–762 (2017). https://doi.org/10.1038/nrclinonc.2017.141
15. Liu, S., et al.: Multimodal neuroimaging computing: a review of the applications in neuropsychiatric disorders. Brain Inf. **2**(3), 167–180 (2015). https://doi.org/10.1007/s40708-015-0019-x
16. Mai, J.K., Paxinos, G., Voss, T.: Atlas of the Human Brain, 3rd edn. Elsevier Inc, New York (2008)
17. Mallat, S.: Recursive interferometric representations. In: European Signal Processing Conference, pp. 716–720 (2010)
18. Mallat, S.: Group invariant scattering. Commun. Pur. Appl. Math. **65**(10), 1331–1398 (2012)
19. Miller, M.I., Tward, D., Trouv'e, A.: Hierarchical computational anatomy: unifying the molecular to tissue continuum via measure representations of the brain. bioRxiv (2021). https://doi.org/10.1101/2021.04.19.440540
20. Miller, M.I., Tward, D.J., Trouve, A.: Coarse-to-fine Hamiltonian dynamics of hierarchical flows in computational anatomy. In: Proceedings of the IEEE/CVF Conference on Computer Vision and Pattern Recognition (CVPR) Workshops (2020)
21. Pantanowitz, L., Sharma, A., Carter, A.B., Kurc, T., Sussman, A., Saltz, J.: Twenty years of digital pathology: an overview of the road travelled, what is on the horizon, and the emergence of vendor-neutral archives. J. Pathol. Inform. **9**(1), 40 (2018). https://doi.org/10.4103/jpi.jpi_69_18
22. Pichat, J., Iglesias, J.E., Yousry, T., Ourselin, S., Modat, M.: A survey of methods for 3D histology reconstruction. Med. Image Anal. **46**, 73–105 (2018). https://doi.org/10.1016/j.media.2018.02.004

23. Ronneberger, O., Fischer, P., Brox, T.: U-Net: Convolutional Networks for Biomedical Image Segmentation. In: Navab, N., Hornegger, J., Wells, W.M., Frangi, A.F. (eds.) MICCAI 2015. LNCS, vol. 9351, pp. 234–241. Springer, Cham (2015). https://doi.org/10.1007/978-3-319-24574-4_28
24. SIfre, L., Mallat, S.: Rigid-motion scattering for texture classification (2014)
25. Tang, Z., Chuang, K.V., DeCarli, C., Jin, L.W., Beckett, L., Keiser, M.J., Dugger, B.N.: Interpretable classification of Alzheimer's disease pathologies with a convolutional neural network pipeline. Nat. Comm. **10**(1), 1–14 (2019). https://doi.org/10.1038/s41467-019-10212-1
26. Tward, D., et al.: Diffeomorphic registration with intensity transformation and missing data: application to 3D digital pathology of Alzheimer's disease. Front. Neurosci. **14**, 1–18 (2020). https://doi.org/10.3389/fnins.2020.00052
27. Tward, D., Li, X., Huo, B., Lee, B., Mitra, P., Miller, M.: 3D mapping of serial histology sections with anomalies using a novel robust deformable registration algorithm. In: Zhu, D., et al. (eds.) MBIA/MFCA -2019. LNCS, vol. 11846, pp. 162–173. Springer, Cham (2019). https://doi.org/10.1007/978-3-030-33226-6_18
28. Xia, C., Babcock, H.P., Moffitt, J.R., Zhuang, X.: Multiplexed detection of RNA using MERFISH and branched DNA amplification. Sci. Rep. **9**(1) (2019). https://doi.org/10.1038/s41598-019-43943-8
29. Yushkevich, P.A., et al.: 3D mapping of tau neurofibrillary tangle pathology in the human medial temporal lobe. In: 2020 IEEE 17th International Symposium on Biomedical Imaging (ISBI), pp. 1312–1316 (2020). https://doi.org/10.1109/ISBI45749.2020.9098462

Multi-scale Hybrid Transformer Networks: Application to Prostate Disease Classification

Ainkaran Santhirasekaram[1,2,3(✉)], Karen Pinto[3], Mathias Winkler[3],
Eric Aboagye[1], Ben Glocker[2], and Andrea Rockall[1,3]

[1] Department of Surgery and Cancer, Imperial College London, London, UK
a.santhirasekaram19@imperial.ac.uk
[2] Department of Computing, Imperial College London, London, UK
[3] Imperial College Healthcare Trust, London, UK

Abstract. Automated disease classification could significantly improve the accuracy of prostate cancer diagnosis on MRI, which is a difficult task even for trained experts. Convolutional neural networks (CNNs) have shown some promising results for disease classification on multi-parametric MRI. However, CNNs struggle to extract robust global features about the anatomy which may provide important contextual information for further improving classification accuracy. Here, we propose a novel multi-scale hybrid CNN/transformer architecture with the ability of better contextualising local features at different scales. In our application, we found this to significantly improve performance compared to using CNNs. Classification accuracy is even further improved with a stacked ensemble yielding promising results for binary classification of prostate lesions into clinically significant or non-significant.

Keywords: Prostate cancer · Convolutional Neural Network · Transformer

1 Introduction

Multi-parametric MRI differentiates non-significant from significant cancers with high accuracy [6]. The scoring system for prostate cancer classification, however, is still somewhat subjective with a reported false positive rate of 30–40% [4]. Therefore there is great interest and clinical need for more objective, automated classification methods for prostate lesions with the goal to improve diagnostic accuracy [7]. There have been various approaches for automated methods based on hand-crafted, quantitative features (radiomics) such as textural and statistical measures that are extracted from regions of interest and used in a machine learning classifier [16]. Textural features have already shown to be beneficial to identify significant prostate cancer [13,19]. Convolutional neural networks are now the most popular approach for automating prostate disease classification on MRI with some promising results [2,7,12,17,21,23]. Due to weight-sharing, the resulting translational

© Springer Nature Switzerland AG 2021
T. Syeda-Mahmood et al. (Eds.): ML-CDS 2021, LNCS 13050, pp. 12–21, 2021.
https://doi.org/10.1007/978-3-030-89847-2_2

invariance and the relatively small receptive field of shallow CNNs, the extracted features tend to capture mostly local information [1]. One has to build much deeper CNNs, combined with down-sampling and multi-scale processing to extract more global information. This problem is relevant to the classification of prostate cancer where global anatomical information and the relationship between different features is important. More recently, the vision transformer has shown competitive performance with convolutional neural networks on natural image classification tasks when pre-trained on large-scale datasets such as ImageNet [20]. However, while transformers can extract better global features by leveraging the power of self-attention for modelling long-range dependencies within the data, the direct tokenisation of patches from an image makes it more difficult to extract more fine low level features while also ignoring locality unlike in deep CNNs. Yet, we know local pixels are highly correlated and this lack of inductive bias means Visual Transformer need a large scale dataset to compete with deep CNN's. It has also been shown that self-attention in the initial layers of a model can learn similar features to convolutions [8]. This shows that the inductive biases imposed by CNNs is appropriate and helpful for feature extraction. Therefore, the use of transformer-only models such as the vision transformer on smaller medical imaging data-sets seems limited and an intriguing approach would be to combine the best of both worlds, giving rise to multi-scale hybrid CNN/transformer networks.

1.1 Contribution

We devise a deep learning approach capable of contextualising local features of prostate lesions through a novel hybrid CNN/transformer architecture with the aim to improve classification of prostate lesion into clinically significant and non-significant. The first stage of our architecture extracts features in a shallow multi-resolution pathway CNN. Each scale extracts CNN based features ranging from more fine grained textural features in the high resolution pathway to more coarse global information from the larger receptive field in the low resolution pathway. Instead of extending the depth of the CNN and using fully connected layers to combine the features within and across different scales, we leverage the powerful self-attention mechanism of the transformer to do this. We specifically use a transformer architecture which takes as input the feature maps from the CNN pathways which we hypothesise will learn better contextualised features to build a richer representation of the input lesion. Our model demonstrates excellent classification accuracy and outperforms a CNN only based approach. We finally further improve performance through a stacked ensemble of our model to outperform other baseline models in the ProstateX challenge [15].

2   Methods

Our proposed model (Fig. 1) has two stages. Stage 1 consists of 3 parallel pathways, each with a different resolution input. There is no weight-sharing between parallel pathways, so each learns discriminative features for a specific resolution.

We use $5 \times 5 \times 3$ convolutional filters in the first layer followed by 2D max pooling in order to account for the anisotropic nature of the input patches. This is followed by residual block layers using $3 \times 3 \times 3$ convolutions and 3D max pooling. We also used grouped convolutions of size 4, which we found to have better performance than single-grouped convolutions during cross-validation for hyper-parameter tuning [22]. The final 128 features maps for each resolution pathway are concatenated to form a stack of 384 feature maps of size $8 \times 8 \times 4$ for stage 1. Each feature map is flattened into a one-dimensional vector forming the input for stage 2 of our model.

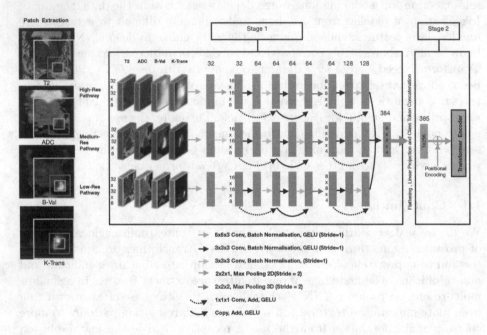

Fig. 1. Proposed model for prostate disease classification. The first stage is the feature extractor for 3 resolution pathways. The second stage is the transformer encoder.

Stage 2 of the model (Fig. 1) involves a linear transformation of the flattened feature maps to an embedding space of dimension 256. A random vector of size 1×256 denoted as the classification token is concatenated to the embedding matrix to learn an image level representation of the feature maps through self-attention. The embedding matrix is of size 385×256. We encode position p using a combination of sine and cosine waves as used in the vision transformer [9]. Each positional encoding i is a vector of $1 \times d$, where d is the embedding dimension with each element in the vector denoted with j. The formulation is described in Eq. 1 below.

$$p_{i,j} = \left\{ \begin{array}{l} \sin(\dfrac{i}{1000^{\frac{j}{d}}}, \text{ if j is even.} \\ \cos(\dfrac{i}{1000^{\frac{j-1}{d}}}, \text{ if j is odd.} \end{array} \right\} \tag{1}$$

We sum the positional encodings to the embeddings to form new embeddings as input into the transformer encoder visualised in Fig. 2. The first part of the encoder consists of layer normalisation followed by a linear transformation of the embedding to Query (Q), Key (K) and value (V) matrices. The Q, K and V matrices are logically split by the number of heads (h) to be of dimension $385 \times 256/h$. Multi-headed self attention is calculated using scaled dot product attention as:

$$head_i = softmax(\frac{Q \times K^T}{\sqrt{256/h}}) \times V \qquad (2)$$

$$Q = M \in \mathcal{R}^{385 \times 256/h}, K = M \in \mathcal{R}^{385 \times 256/h}, V = M \in \mathcal{R}^{385 \times 256/h}$$

$$MultiHead(Q, K, V) = Concat(Head_1, Head_2...Head_h) \times W_0 \qquad (3)$$

$$MultiHead(Q, K, V) = M \in \mathcal{R}^{256 \times 385}, W_0 = M \in \mathcal{R}^{256 \times 256}$$

The different heads are concatenated and undergo linear transformation (Eq. 3). The next stage is a multi-layer perceptron (MLP) with two fully connected layers (Fig. 2). The transformer encoder network is repeated L (layers) times. Residual connections are incorporated to aid with gradient flow. We use Gaussian error linear unit (GELU) activation for both stages [11] and a dropout rate of 0.2 after every dense layer except for linear transformation of Q, K, V in stage 2 of the model. During 5-fold cross validation we observe the optimum number of heads, layers and MLP hidden dimension to be 8, 8 and 1024 respectively. Finally, the learnt classification token vector is consumed by an MLP classifier which consists of two fully connected layers with a hidden dimension of 1024 (Fig. 2). Stage 1 and 2 of the model are trained end-to-end.

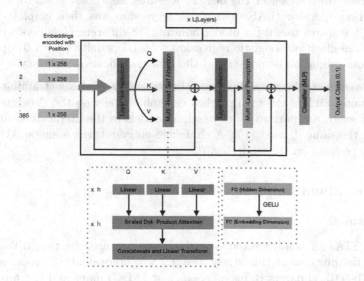

Fig. 2. Transformer encoder architecture with L layers and h heads for multi headed self attention. An MLP is used as the classifier after L layers of the Transformer encoder

2.1 Model Comparison

CNNs: We compare with a number of models using convolutions only. Firstly, we train the three resolution pathways in stage 1 of our model separately followed by 2 fully connected layers of size 4096 and 512 with dropout (0.5) to form 3 separately trained CNN models without masking. We then train a new high and medium resolution CNN with tumour masking followed by a new low resolution CNN with whole prostate masking as its goal is to extract prostate anatomical as well as lesion information. After evaluating the effect of masking on classification performance, we trained two multi-resolution CNNs (high and medium resolution vs all 3 resolutions). We replace the second stage with 2 fully connected layers of size 8192 and 1024 with dropout (0.5) for both multi-resolution CNNs.

Model Ensemble: To improve classification accuracy of our proposed model, We trained an ensemble of our multi-Scale hybrid transformer by varying the number of epochs during training (14–16 epochs), number of layers (6–9 Layers) and the MLP hidden dimension (1024 and 1280) to produce 24 trained models for each training fold in 5-fold cross-validation. We employ a stacked ensemble method and use the class output probability from each model as input to train a logistic regression model with L2 penalty ($\lambda = 1.0$) to predict the class outputs.

Radiomics: A total of 2724 features were extracted using TexLab2.0; a Radiomics analysis software [3]. The tumour segmentations were used for masking. Various feature selection methods were explored during 5 fold cross-validation. The best feature selection process firstly involved removing highly correlated features (correlation coefficient > 0.95) using Pearson Correlation [5]. Secondly, uni-variate feature selection with Analysis of variance (ANOVA) analysis was performed to select the best 20 features [18]. Least absolute shrinkage and selection operator (LASSO) logistic regression was then employed which identified 6 features useful for model building. 18 different classifiers were evaluated for classification. Logistic regression with L2 penalty ($\lambda = 1.0$) optimised during cross-validation demonstrated the best classification accuracy.

Other Baselines: We trained only on the ProstateX challenge training dataset and therefore validate our final stacked ensemble model on the ProstateX challenge test set to compare to other baseline models in the literature trained and tested on the same dataset [15]. A challenge entry returns a single AUC-ROC value and position on the leaderboard.

3 Experiments and Results

3.1 Dataset

The PROSTATEx challenge dataset which was acquired on two, 3 T scanners [15]. For the purpose of this study, the T2 weighted axial, diffusion weighted imaging (b-800), apparent diffusion coefficient (ADC) maps and K-trans images were used. The dataset consists of 330 pre-selected lesions [15]. Clinically significant lesions are classed as Gleason grade group 2 and above. Non-biopsied lesions were considered non-significant as ground truth and histology results were used as ground truth for biopsied lesions. The dataset is imbalanced with only 23% of lesions labelled as significant.

Segmentation of each lesion and prostate was performed by a Radiologist using ITKsnap [24]. The lesion segmentation was performed on the T2 axial, ADC, b-800 and K-trans. All patients had a lesion visible on at least one sequence which were segmented manually. The sequence with the maximum lesion volume (from sequences with the lesion visible) is mapped to the sequences where a lesion is not visible. Whole prostate segmentation was performed on the T2 axial only.

3.2 Pre-processing and Augmentation

We resample to form 3 sets of images: high resolution (0.5 mm × 0.5 mm × 1.5 mm), medium resolution (1.0 mm × 1.0 mm × 3.0 mm) and low resolution (2.0 mm × 2.0 mm × 4.0 mm). Cubic B-spline interpolation is used for resampling of the MR images. Nearest-neighbour is used for resampling of the mask. The high resolution images were used for radiomics input. The ADC, b800 and K-trans images were then registered with the T2 axial images using affine transformation. We then mask out irrelevant background areas that are further away from the boundary of the prostate as only nearby extra-prostatic regions are assumed to be valuable for tumour classification. We do this by using the boundaries of the whole prostate segmentation mask to form an initial bounding box for each slice around the prostate. The bounding box is then extended to include more extra-prostatic region by adding an optimal length to the width and height of the bounding box defined as 10 divided by the resolution (mm) in the axial plane.

Patches were then extracted for each resolution. For, high and medium resolution images, $32 \times 32 \times 8$ patches were extracted centred on the lesion. In the low resolution images, the goal is to capture as much of the prostate in the region of interest (ROI). Therefore we limit the area outside of the prostate, by centering a $32 \times 32 \times 8$ patch on a point equidistant between the lesion and whole prostate centre. All patches are then processed with and without masking as described in Sect. 2.1. Finally, we re-scaled the intensities between 0 and 1 for normalisation.

We augment the significant class to handle class imbalance through vertical or horizontal flipping followed by random rotations between −90 and 90° and random translation of 0 to 10 mm.

3.3 Training

Despite accounting for class imbalance using augmentation. This would not completely account for natural variations of significant tumour appearance. Therefore we use a weighted binary cross entropy loss function (Eq. 4) with the weighting parameters fine tuned during cross validation. This was applied to all models.

$$WeightedCrossEntropy = -0.6log(p) - 0.4log(1 - p) \qquad (4)$$

All model were implemented using PyTorch. Experiments were run on three NVIDIA Geforce RTX 2080 GPUs. Stratified (using label) 5 fold cross-validation is used for model training/validation and to optimise the number of epochs, batch size, learning rate, weight decay, loss function weightings, dropout rate, the transformer encoder hidden dimension and number of CNN/transformer encoder layers. The model weights are initialised with Kaiming initialisation [10]. The CNN based models use Adam optimisation with a base learning rate of 0.0001 [14] and weight decay of 0.001 for all models. The hybrid model uses Adam optimisation with weight decay (0.001) and cosine annealing (base learning rate: 0.001) as a learning rate scheduler. We use a batch size of 40 for training. The single and two resolution CNNs were trained for 10 epochs. The multi-resolution CNN and our multi-Scale hybrid transformer were trained for 12 and 15 epochs, respectively.

For testing, our final proposed stacked ensemble of multi-scale hybrid transformers is trained on the entire dataset and submitted for external validation on the ProstateX test set.

3.4   Results

Table 1. Mean and standard error on 5-fold cross-validation for metrics comparing all trained models. Best result for each metric is highlighted in bold.

	Accuracy	Specificity	Precision	Recall
High Res CNN (no mask)	0.841 ± 0.009	0.811 ± 0.025	0.826 ± 0.020	0.866 ± 0.015
High Res CNN (mask)	0.840 ± 0.009	0.813 ± 0.015	0.828 ± 0.056	0.868 ± 0.022
Medium Res CNN (no mask)	0.810 ± 0.001	0.794 ± 0.019	0.802 ± 0.004	0.866 ± 0.043
Medium Res CNN (mask)	0.818 ± 0.043	0.800 ± 0.019	0.803 ± 0.035	0.857 ± 0.045
Low Res CNN (no mask)	0.771 ± 0.037	0.751 ± 0.039	0.739 ± 0.070	0.788 ± 0.053
Low Res CNN (mask)	0.764 ± 0.012	0.742 ± 0.058	0.752 ± 0.028	0.782 ± 0.043
Two Res CNN (no mask)	0.875 ± 0.009	0.854 ± 0.021	0.860 ± 0.006	0.888 ± 0.019
Three Res CNN (no mask)	0.883 ± 0.008	0.865 ± 0.028	0.868 ± 0.009	0.895 ± 0.022
Radiomics	0.775 ± 0.053	0.751 ± 0.055	0.753 ± 0.048	0.799 ± 0.040
Our model (no mask)	0.900 ± 0.018	0.899 ± 0.024	0.879 ± 0.014	0.918 ± 0.017
Our model ensemble	$\mathbf{0.944 \pm 0.013}$	$\mathbf{0.927 \pm 0.023}$	$\mathbf{0.933 \pm 0.009}$	$\mathbf{0.959 \pm 0.022}$

We observe slightly improved performance from not using a mask for all three single resolution CNN models (Table 1). This suggests that the prostate area outside the volume of the tumour and nearby area outside the prostate itself provides useful information for classification. We therefore did not use whole prostate and tumour masks for the multi-resolution CNNs and our model.

We also find that increasing the input resolution of the CNN improves classification performance (Table 1). This is most likely due to class imbalance as the ROI increases which is more pronounced after masking. This is also likely due to the loss of fine features important for classification at coarser resolutions. However,

we find the two resolution CNN demonstrates improved performance in evaluation metrics which is further enhanced with the three resolution CNN (Table 1).
This demonstrates the importance of combining local and global features to learn
better contextual information of the tumour lesion using the lower resolution pathways to provide more informative anatomical localisation of the tumour region.
Our model outperforms all CNN only models in all metrics (Table 1). This shows
our model improves CNN performance by harnessing self-attention in the transformer to extract better contextual features by learning important relationships
between the features maps extracted in each CNN pathway. We also demonstrate
significantly better performance of using multi-scale CNNs and our model compared to the radiomics approach (Table 1) highlighting the benefit of feature learning. A stacked ensemble of our model leads to overall best performance on all evaluation metrics (Table 1) and achieves an average AUC of 0.95 during 5- fold cross-
validation (Fig. 3). Our model ensemble achieves an AUC of 0.94 and 3rd place on
the leader-board for the ProstateX challenge test set.

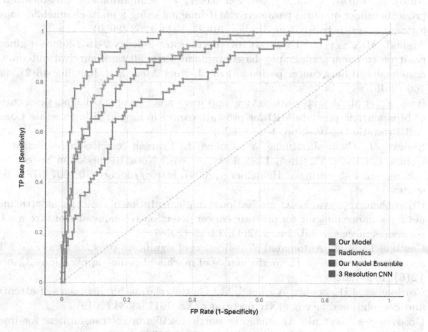

Fig. 3. OC curves from the merged predictions of each fold comparing our model and
model ensemble to radiomics and the three resolution CNN.

4   Conclusion

We demonstrate the importance of extracting contextual information of the
tumour region in regards to its anatomical location and extension. We propose
a novel multi-scale hybrid CNN/transformer network with the ability to extract

richer contextualised features to build stronger representations which significantly improves prostate disease classification in all evaluation metrics compared to radiomics and multi-resolution CNNs. We believe our novel transformer-based approach could be appealing for many other disease classification tasks where the contextualisation of fine-detailed local features is important. This will be explored in future work.

Acknowledgments. This work was supported and funded by Cancer Research UK (CRUK) (C309/A28804).

References

1. Albawi, S., Mohammed, T.A., Al-Zawi, S.: Understanding of a convolutional neural network. In: 2017 International Conference on Engineering and Technology (ICET), pp. 1–6. IEEE (2017)
2. Aldoj, N., Lukas, S., Dewey, M., Penzkofer, T.: Semi-automatic classification of prostate cancer on multi-parametric MR imaging using a multi-channel 3d convolutional neural network. Eur. Radiol. **30**(2), 1243–1253 (2020)
3. Arshad, M.A., et al.: Discovery of pre-therapy 2-deoxy-2-18 f-fluoro-d-glucose positron emission tomography-based radiomics classifiers of survival outcome in non-small-cell lung cancer patients. Eur. J. Nucl. Med. Mol. Imaging **46**(2), 455–466 (2019)
4. Bass, E., et al.: A systematic review and meta-analysis of the diagnostic accuracy of biparametric prostate MRI for prostate cancer in men at risk. Prostate Cancer and Prostatic Diseases, pp. 1–16 (2020)
5. Benesty, J., Chen, J., Huang, Y., Cohen, I.: Pearson correlation coefficient. In: Cohen, I., Huang, Y., Chen, J., Benesty, J. (eds.) Noise Reduction in Speech Processing, pp. 1–4. Springer, Heidelberg (2009). https://doi.org/10.1007/978-3-642-00296-0
6. Brizmohun Appayya, M., et al.: National implementation of multi-parametric magnetic resonance imaging for prostate cancer detection-recommendations from a UK consensus meeting. BJU Int. **122**(1), 13–25 (2018)
7. Castillo, T., et al.: Automated classification of significant prostate cancer on MRI: a systematic review on the performance of machine learning applications. Cancers **12**(6), 1606 (2020)
8. Cordonnier, J.B., Loukas, A., Jaggi, M.: On the relationship between self-attention and convolutional layers. arXiv preprint arXiv:1911.03584 (2019)
9. Dosovitskiy, A., et al.: An image is worth 16x16 words: transformers for image recognition at scale. arXiv preprint arXiv:2010.11929 (2020)
10. He, K., Zhang, X., Ren, S., Sun, J.: Delving deep into rectifiers: surpassing human-level performance on ImageNet classification. In: Proceedings of the IEEE International Conference on Computer Vision, pp. 1026–1034 (2015)
11. Hendrycks, D., Gimpel, K.: Gaussian error linear units (GELUs). arXiv preprint arXiv:1606.08415 (2016)
12. Ishioka, J., et al.: Computer-aided diagnosis of prostate cancer on magnetic resonance imaging using a convolutional neural network algorithm. BJU Int. **122**(3), 411–417 (2018)

13. Khalvati, F., Wong, A., Haider, M.A.: Automated prostate cancer detection via comprehensive multi-parametric magnetic resonance imaging texture feature models. BMC Med. Imaging **15**(1), 1–14 (2015)
14. Kingma, D.P., Ba, J.: Adam: a method for stochastic optimization. arXiv preprint arXiv:1412.6980 (2014)
15. Litjens, G., Debats, O., Barentsz, J., Karssemeijer, N., Huisman, H.: Computer-aided detection of prostate cancer in MRI. IEEE Trans. Med. Imaging **33**(5), 1083–1092 (2014)
16. Rizzo, S., et al.: Radiomics: the facts and the challenges of image analysis. Eur. Radiol. Exp. **2**(1), 1–8 (2018). https://doi.org/10.1186/s41747-018-0068-z
17. Song, Y., et al.: Computer-aided diagnosis of prostate cancer using a deep convolutional neural network from multiparametric MRI. J. Magn. Reson. Imaging **48**(6), 1570–1577 (2018)
18. St, L., Wold, S., et al.: Analysis of variance (ANOVA). Chemom. Intell. Lab. Syst. **6**(4), 259–272 (1989)
19. Stoyanova, R., et al.: Prostate cancer radiomics and the promise of radiogenomics. Transl. Cancer Res. **5**(4), 432 (2016)
20. Vaswani, A., et al.: Attention is all you need. arXiv preprint arXiv:1706.03762 (2017)
21. Wang, Z., Liu, C., Cheng, D., Wang, L., Yang, X., Cheng, K.T.: Automated detection of clinically significant prostate cancer in MP-MRI images based on an end-to-end deep neural network. IEEE Trans. Med. Imaging **37**(5), 1127–1139 (2018)
22. Xie, S., Girshick, R., Dollar, P., Tu, Z., He, K.: Aggregated residual transformations for deep neural networks. In: Proceedings of the IEEE Conference on Computer Vision and Pattern Recognition (CVPR), July 2017
23. Yang, X., et al.: Co-trained convolutional neural networks for automated detection of prostate cancer in multi-parametric MRI. Med. Image Anal. **42**, 212–227 (2017)
24. Yushkevich, P.A., Gao, Y., Gerig, G.: ITK-snap: an interactive tool for semi-automatic segmentation of multi-modality biomedical images. In: 2016 38th Annual International Conference of the IEEE Engineering in Medicine and Biology Society (EMBC), pp. 3342–3345. IEEE (2016)

Predicting Treatment Response in Prostate Cancer Patients Based on Multimodal PET/CT for Clinical Decision Support

Sobhan Moazemi[1,2]($\boxtimes$) (iD), Markus Essler[2], Thomas Schultz[1,3] (iD),
and Ralph A. Bundschuh[2]

[1] Department of Computer Science, University of Bonn, 53115 Bonn, Germany
schultz@cs.uni-bonn.de
[2] Department of Nuclear Medicine, University Hospital Bonn, 53127 Bonn, Germany
{s.moazemi,markus.essler,ralph.bundschuh}@ukbonn.de
[3] Bonn-Aachen International Center for Information Technology (BIT),
53115 Bonn, Germany

Abstract. Clinical decision support systems (CDSSs) have gained critical importance in clinical practice and research. Machine learning (ML) and deep learning methods are widely applied in CDSSs to provide diagnostic and prognostic assistance in oncological studies. Taking prostate cancer (PCa) as an example, true segmentation of pathological uptake and prediction of treatment outcome taking advantage of radiomics features extracted from prostate-specific membrane antigen-positron emission tomography/computed tomography (PSMA-PET/CT) were the main objectives of this study. Thus, we aimed at providing an automated clinical decision support tool to assist physicians. To this end, a multi-channel deep neural network inspired by U-Net architecture is trained and fit to automatically segment pathological uptake in multimodal whole-body baseline ^{68}Ga-PSMA-PET/CT scans. Moreover, state-of-the-art ML methods are applied to radiomics features extracted from the predicted U-Net masks to identify responders to ^{177}Lu-PSMA treatment. To investigate the performance of the methods, 2067 pathological hotspots annotated in a retrospective cohort of 100 PCa patients are applied after subdividing to train and test cohorts. For the automated segmentation task, we achieved 0.88 test precision, 0.77 recall, and 0.82 Dice. For predicting responders, we achieved 0.73 area under the curve (AUC), 0.81 sensitivity, and 0.58 specificity on the test cohort. As a result, the facilitated automated decision support tool has shown its potential to serve as an assistant for patient screening for ^{177}Lu-PSMA therapy.

Keywords: Clinical decision support system · Machine learning · Deep learning · Multimodal Imaging · Positron emission tomography · Computed tomography · Prostate cancer

© Springer Nature Switzerland AG 2021
T. Syeda-Mahmood et al. (Eds.): ML-CDS 2021, LNCS 13050, pp. 22–35, 2021.
https://doi.org/10.1007/978-3-030-89847-2_3

1   Introduction

Computer-aided diagnosis (CAD) has been used extensively to assist physicians and researchers in a variety of fields including oncology. Prostate cancer endangers men's health as the fifth cancer disease to cause mortality in the world [1]. To assess the disease stage as well as to monitor the treatment progress, PET/CT scans are commonly used. PET/CTs are multimodal medical imaging techniques which are widely used for different cancer diseases [2–4,12]. On the one hand, PET scans outline differences in functional activities of different tissues. On the other hand, CT scans provide high-resolution spatial and anatomical information of the tissues. Thus together, PET and CT provide both functional and anatomical information to locate malignancies.

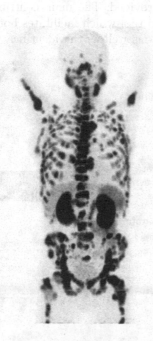 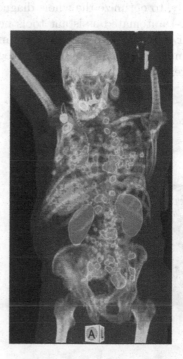

Fig. 1. An example of multimodal imaging for prostate cancer management. Left: the positron emission tomography (PET), right: the overlaid PET/computed tomography (PET/CT). The red uptake in the right panel includes both pathological and physiological uptake.

In clinical routine, nuclear medicine (NM) physicians often spend an enormous amount of time to analyze medical imaging modalities to locate cancerous tissues and to stage the disease. Moreover, considering the treatment expenses, avoiding unnecessary treatment would be of critical importance. Thus, on the one hand, clinical decision support tools can reduce the time and effort required by the domain experts, and on the other hand, they could help with identifying the patients who might not benefit from the therapy in a timely manner.

In order to make diagnostic and prognostic decisions based on PET/CT scans, several issues need to be taken care of. First of all, PET and CT scans, even when taken with the same scanner, are produced in different resolutions and often in different coordination systems. Second of all, as a non-invasive diagnostic method, true segmentation of pathological (i.e., malignant) uptake is of critical importance. Last but not least, the required time and effort to conduct decisive actions need to be minimized. To tackle the first challenge, co-registration and resampling of the two modalities and bringing them to the same coordinate system have to be conducted. Furthermore, for the segmentation of the malignant tissues, appropriate labeling and annotation tools should be applied. Finally, to optimize the whole diagnostic procedure in terms of time and effort, (semi-) automated assistant tools need to be provided. The main contribution of our methods is to provide such an automated tool which facilitates both the automated segmentation of pathological uptake as well as the identification of non-responders to therapy.

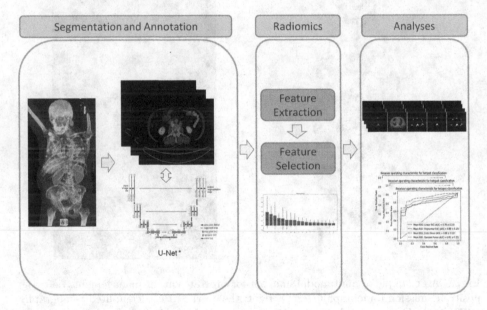

Fig. 2. The high-level outline of the tools. The segmentation and annotation module consists of in-house developed and third party software for automatic and manual segmentation and labeling of the input scans. The radiomics unit utilizes radiomics feature calculation and selection. The analyses module encapsulates the tools used for the visualization of study results (*: the original U-Net architecture as proposed by Ronneberger et al. [8]).

In general, given PET 3D scans, cancerous and metastatic tissues usually feature higher metabolic uptake than benign or normal tissues. However, challenges arise as some organs such as liver and kidneys also associate with high uptake which is considered as not pathological (i.e., physiological). Thus, discriminating pathological uptake becomes more challenging for inexperienced annotators. To cope with this issue, often CT scans complement PET scans as an additional channel to better locate the sites with pathological uptake. Figure 1 illustrates an example of multimodal PET/CT scans, denoting the similarities between pathological and physiological uptakes.

As conventional segmentation methods, thresholding based algorithms are applied to PET scans, leveraging fixed and adaptive thresholds such as 40% of maximum standardized uptake values (40%-SUV$_{MAX}$). However, inconsistencies in scanning resolutions and protocols from scanner to scanner and from center to center expose limitations to thresholding based approaches. To address this issue, automated segmentation leveraging deep neural networks have been beneficial in many diagnostics fields such as oncology [5], drusen segmentation [6] and computational neuroscience [7]. Since first time proposed by Ronneberger et al. [8], U-Net based architectures have been widely applied in medical domain for different segmentation tasks [9,10]. Liu et al. [11] have conducted a survey on U-shaped networks used for medical image segmentation.

As the next consecutive step towards diagnosis and therapy response prediction, deep and supervised ML methods are commonly used in combination with radiomics analyses in clinical research [12–14]. The term radiomics denotes the procedure of extracting numerical quantities out of medical imaging data in terms of two or three dimensional (2D or 3D) intensity, shape, or texture based features which characterize tumors and other hotspots.

In this manuscript, we introduce a clinical decision support tool consisting of an automated segmentation unit inspired by U-Net architecture and supervised ML classifiers to distinguish responders and non-responders to ^{177}Lu-PSMA as a routine therapy procedure to treat patients diagnosed with prostate cancer. The presented automated tools facilitate the management of PCa patients based on baseline ^{68}Ga-PSMA-PET/CT scans. To end up with the ultimate decision support solution, several steps are taken: 1) preprocess and resample PET and CT input Dicom [15] images, 2) annotate the input images to provide ground truth (GT) labels for ML and deep learning pipelines, 3) automatically segment the pathological uptake using deep neural networks (NNs) and extract radiomics features based on the predicted masks, 4) utilize supervised ML classifiers to predict the responders to therapy based on the radiomics features extracted from the PET/CT findings.

For this study, a cohort of 100 PCa patients have been retrospectively analyzed and subdivided to training and test cohorts as detailed in Sect. 2.2. The training cohort is used for training and fitting the automated segmentation pipeline and for training and hyperparameter tuning of the supervised ML classifiers as used for therapy response prediction. We took advantage of TensorFlow [16] and Keras [17] libraries to customize our U-Net based segmentation model based on multimodal PET and CT channels to predict masks based on manually annotated GT masks using third party software. Dice coefficient [18] is applied to quantify the amount of agreement between the predicted masks and the given GT masks. As a result, the superiority of combination of PET and CT channels compared to masks predicted using only PET channel is shown. In the next step, PyRadiomics library [19] is used to calculate the radiomics features for PET and CT modalities based on the pathological uptake predicted by the implemented PET-CT-U-Net. In the final step, ML classifiers are applied to the calculated radiomics features to predict responders to therapy. This step includes recursive feature elimination (RFE) technique [20] to identify the most relevant features for the classification problem.

To the best of our knowledge, so far, no other automated clinical decision support tool based on deep learning methods is presented for the management of prostate cancer patients using baseline ^{68}Ga-PSMA-PET/CT findings.

2   Methods

2.1   Pipeline Overview

The study pipeline consists of several modules including multimodal PET/CT resampling and visualization tool box, U-Net based segmentation, radiomics feature extraction and selection, and prognosis which together serve as a clinical decision support system (CDSS) for the management of PCa patients. Figure 2 gives a high-level outline of the tools. The whole pipeline is represented as three consecutive building blocks: 1) segmentation and annotation, 2) radiomics, and 3) analyses.

The segmentation and annotation block consists of various assets facilitating manual and automated segmentation and annotation of multimodal PET/CT images. For instance, InterView FUSION [21] is used for the manual delineation of hotspots while an in-house developed deep segmentation network based on U-Net [8] is used for automated segmentation of pathological uptake. The radiomics block consists of feature extraction and selection tools. Finally, the analyses block covers the tools and diagrams aiming at representing the end-user level analyses and insights.

Table 1. The summary of the clinical information of the patients cohort (PSA: prostate specific antigen).

	Age [years]	Gleason score	PSA [ng/ml]
Minimum	48	6	0.25
Maximum	87	10	5910
Average	70.40	8.32	461.57

2.2   Dataset and Ground Truth Annotation

For this study, a cohort of 100 patients (67 responders vs 33 non-responders) who had been diagnosed with advanced prostate carcinoma and had been selected for ^{177}Lu-PSMA therapy has been analyzed retrospectively. The patients' ages ranged from 48 to 87 years. Table 1 gives an overview of the clinical information of the cohort. A Biograph 2 PET/CT system (Siemens Medical Solutions, Erlangen, Germany) was used for image acquisition and reconstruction. A random subset of 61 subjects (40 responders vs 21 non-responders) is used for training, while the rest of the patients (39 patients including 27 responders and 12 non-responders) have served as test cohort. All patients gave written and informed consent to the imaging procedure and for anonymized evaluation of their data. Due to the retrospective character of the data analysis, an ethical statement was waived by the institutional review board according to the professional regulations of the medical board of the state of North-Rhine Westphalia, Germany.

To provide ground truth (GT) labels for the ML based pipeline, a team of NM physicians supervised by a highly qualified NM expert with 7 years of practical experience in PET/CT diagnosis analyzed and annotated the whole dataset using InterView FUSION. In a previous study [27], we have analyzed the inter-observer variability aspects of GT annotation. The dataset is annotated in a slice-based manner. Thus, each pathological hotspot was delineated as several consecutive 2D regions of interest (RoIs) in subsequent slices, together forming the volumes of interest (VoIs). On average, 20 pathological hotspots have been identified for each patient (in total, 2067 pathological hotspots). The pathological hotspots include primary uptake in prostate as well as metastatic uptake in other organs such as bone and lymph nodes.

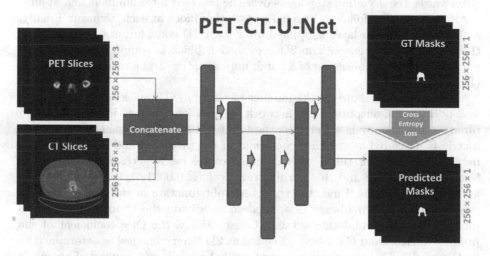

Fig. 3. The simplified schematic of the implemented multimodal U-Net based segmentation network (PET-CT-U-Net). PET and CT slices are processed as separate channels. Two alternative models are applied: PET only, and PET/CT. Weighted binary cross-entropy serves as the loss function.

2.3   Automated Segmentation

The U-Net based segmentation unit, inspired by a single channel U-Net model implemented for polyp segmentation [22], is a multi-channel network which takes resampled images from original Dicom format PET and CT images as 2D sliced input. In addition, the 2D sliced manually delineated GT masks are used as GT labels. Also, the network internally creates the threshold based 40%-SUV_{MAX} masks from PET for quantitative as well as qualitative comparison. The segmentation network defines two different models based on input channels: single and dual. The simplified architecture of the automated segmentation network is illustrated in Fig. 3.

In the first training step, the U-Net based model is trained and fit using PET and the combinations of PET and CT. As a result, two alternative models are trained, a model based on only PET and a model based on both PET and CT images respectively. The 40%-SUV_{MAX} masks are used to set a baseline for performance analysis of the segmentation network. The U-Net model, developed in Python V.3.6 and utilized TensorFlow and Keras libraries, consists of encoding and decoding steps connected via a bridge layer. The input image sizes are set to 256×256 and the filter numbers are as follows: 16, 32, 48, 64, 128, 256, 480, 512. Thus, we customize the resolution levels compared to the original U-Net architecture. By looping over the filter numbers and applying a 2D convolution block followed by a 2D max pooling at each iteration, the encoding step is taken. Then the bridge layer comes to action by applying a single 2D convolution block. Afterwards, the decoding step loops over the reversed filter numbers and applies a 2×2 upsampling followed by a 2D convolution block at each iteration. Finally, a sigmoid activation layer is applied after 1×1 2D convolution to end up with the output binary image. The 2D convolution block is composed of two 3×3 convolutions, each consisting of a batch normalization and a rectified linear unit (ReLU) activation.

Due to the imbalance in the number of pixels in GT masks and background and because the outputs of the network are predicted binary images, weighted binary cross entropy is selected as the loss function. The segmentation quality metric is measured as the Dice coefficient of the predicted and GT masks. The networks are further tuned with different values for hyperparameters such as batch size (values: 8 and 16), learning rate (values: 0.0001, 0.001, 0.01, and 0.1), and up to 60 epochs. First, the train_test_split function of the model_selection class of Scikit-Learn library is applied to subdivide the training cohort into interim train and validation subsets. Consecutively, the Dice coefficient of the predicted masks and GT labels (encoded as 2D binary images) is determined to fit the model. This process is repeated until the maximum number of epochs is reached or the early stopping criteria (using TensorFlow's EarlyStopping function with inputs: monitor = validation_loss and patience = 10) are met. Then, the fitted model is used to predict the masks for the held-out test set. Finally, the generated masks will be analyzed quantitatively (using Dice coefficients) as well as qualitatively (by the experienced NM expert). To prepare the input dataset for the therapy response prediction, the best predicted mask is used to calculate radiomics features. To this end, for each patient, the predicted mask is applied to input images to calculate features using PyRadiomics library and end up with patient-specific feature vectors.

2.4   Therapy Response Prediction

Once the radiomics features are generated from the segmentation unit, 6 different ML classifiers (logistic regression [23], support vector machine (SVM) [24] with linear, polynomial and radial basis function (RBF) kernels, extra trees [25] and random forest [26]) are used for the task of prediction of responders to ^{177}Lu-PSMA treatment. We further analyze the radiomics features and apply recursive feature elimination to end up with the most relevant features to the classification problem. Here, the same training cohort of 61 subjects as used for the training of the segmentation network is applied for training and hyperparameter tuning in a cross validation (CV) step. For the CV, stratified KFold is used with 3 folds. In each CV step, the train and validate feature vectors are standardized using MinMax standardization method. For the hyperparameter tuning, grid search is applied for all of the classifiers based on standard possible ranges of each hyperparameter such as C and Gamma for the SVM classifiers and max_depth and min_sample_leaf for the decision tree based classifiers. For example, the hyperparameter C ranged from 2^{-5} to 2^{15} and min_sample_leaf ranged from 1 to 10. In the final step, the performance of each classifier is measured as applied to the feature vectors from the held-out test cohort. The performances are quantified as area under the receiver operating characteristic (ROC) curve (AUC), sensitivity (SE), and specificity (SP). As the baseline for comparison, the classifiers performances are quantified as the GT mask is applied for the calculation of radiomics features using PyRadiomics library.

Table 2. The performances of different U-Net based segmentation models as trained and fit with the training cohort and applied to the test cohort. The performance of 40%-SUV$_{MAX}$ mask has been quantified for comparison. The precision, recall and Dice values are mean and standard deviations over the test subject cohort. (lr: learning rate, acc: accuracy).

Model/Mask	epochs	lr	acc	Dice	Loss	Precision	Recall
40%-SUV$_{MAX}$	–	–	99	39.62 ± 16.6	0.01	38.53 ± 21.38	51.48 ± 19.19
PET (Single)	35	0.001	99	71.51 ± 4.9	0.01	83.63 ± 5.3	63.38 ± 4.8
PET/CT (Dual)	32	0.001	99	82.18 ± 4.7	0.01	88.44 ± 4.8	77.09 ± 5.7

3   Results

3.1   Segmentation

As described in Methods section, we analyzed both singular (i.e. just PET) and multiple (PET + CT) input channels to train and fit our segmentation model.

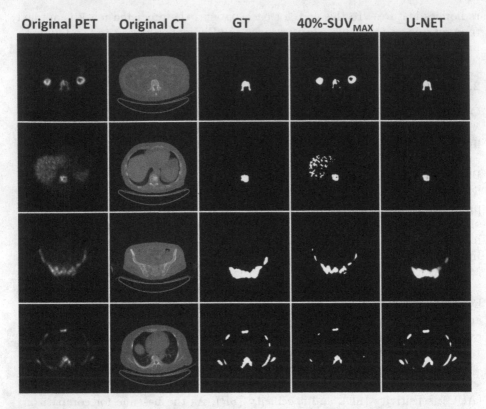

Fig. 4. Example slices of the U-Net based segmentation results. The input PET and CT slices, the ground truth (GT), 40%-SUV$_{MAX}$ PET, and predicted masks are shown. The rows belong to unique 2D slices from arbitrary subjects of the test cohort.

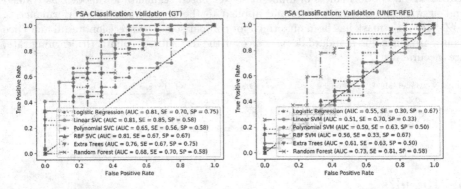

Fig. 5. Receiver operating characteristic (ROC) curves based on GT masks and U-Net predicted masks with feature selection. The 6 classifiers are trained and tuned on the training set and applied to the test set (RBF: radial basis function, RFE: recursive feature elimination, AUC: area under the curve, SE: sensitivity, SP: specificity).

As a result, the best performance in terms of accuracy, precision, recall, and Dice in training and test were observed by the multi-channel model with batch size of 16, 0.99 test accuracy, 0.88 test precision, 0.77 test recall, and 0.82 test Dice. Table 2 compares the achieved performances from the alternative U-Net models and those of 40%-SUV$_{MAX}$ masks. Figure 4 provides a qualitative outline of the segmentation results, comparing original input channels and multi-channel U-Net model prediction. As the results suggest, the U-Net predicted mask performs reasonably well as compared to the GT mask. Moreover, the U-Net prediction outperformed the thresholding based mask, specifically, for the identification of the physiological uptake (e.g., in livers and kidneys) which is considered as one of the challenging tasks for computer based algorithms [27,28]. Furthermore, the proposed automated segmentation performs well in predicting bone metastasis uptakes.

Table 3. The most relevant radiomics features selected by recursive feature elimination (RFE) from both PET and CT modalities. For more information on the radiomics features, refer to [19] (glrm: gray level run length matrix, glszm: gray level size zone matrix).

Feature group	Feature subgroup	Feature name
pet: diagnostics	Image-original	Mean
pet: original	shape	SurfaceVolumeRatio
ct: original	shape	MinorAxisLength
pet: original	firstorder	Energy
pet: original	firstorder	Maximum
pet: original	firstorder	Skewness
pet: original	glrlm	RunEntropy
pet: original	glrlm	RunLengthNonUniformityNormalized
pet: original	glrlm	RunPercentage
ct: original	glrlm	ShortRunEmphasis
ct: original	glszm	SmallAreaEmphasis
ct: original	glszm	SmallAreaLowGrayLevelEmphasis
pet: original	glszm	ZonePercentage
ct: original	glszm	ZonePercentage

3.2   Therapy Response Prediction

To analyze classifiers' performances, radiomics features have been calculated for both GT and U-Net predicted masks. A total of 120 radiomics features including first and higher-order statistics such as minimum and maximum intensity, textural heterogeneity parameters such as entropy and kurtosis, and run and zone-length statistics are calculated for both PET and CT modalities using

PyRadiomics library. For the complete list of the features visit PyRadiomics official documentation [19]. Among all the classifiers, logistic regression performed the best as applied to the radiomics features calculated based on GT masks with AUC = 0.81, SE = 0.70, SP = 0.75 on the test set. As the prediction performances of all the classifiers as applied to U-Net predicted masks were not satisfactory (AUC ranged from 0.41 to 0.55), recursive feature elimination technique has been applied to identify most relevant features for the classification task. Table 3 shows the list of 14 features as selected by RFE method. Taking advantage of feature selection, the classification performances of most of the classifiers have been clearly improved. As previous work [14,27] had suggested, features from both PET and CT modalities contribute to the classification task. The overall best performance belongs to random forest classifier with AUC = 0.73, SE = 0.81, SP = 0.58 as applied to the held-out test cohort. Figure 5 shows receiver operating characteristic (ROC) curves for all of the 6 classifiers as applied to radiomics features calculated based on GT and U-Net predicted masks after applying recursive feature elimination (RFE).

4   Discussion

Facilitating fast and accurate non-invasive diagnosis and prognosis has been the objective of computer-aided diagnosis (CAD) for years. When it comes to the oncological domain, especially in subjects in advanced metastatic stages, CAD systems take over the histopathological analyses in many clinical practices. This is globally justified as taking multiple biopsies from patients is ethically questionable. However, the procedure of manual delineation of the malignant tissues using established tools such as InterView FUSION is considered time consuming and attention intensive. Therefore, the first goal of this study was to develop an automated segmentation tool for multimodal PET/CT scans.

We retrospectively analyzed 2067 pathological hotspots from 100 PCa patients (on average, 20 pathological hotspots per patient). As shown in Results section, our U-Net based multi-channel segmentation network predicts the pathological masks with a high accuracy. Particularly, we showed that including the PET and CT modalities as multiple channels outperforms predictions of the U-Net model as trained only using the original PET channel. Also, the qualitative analyses revealed that the multi-channel U-Net prediction is superior in discriminating non-pathological uptake in liver and kidneys compared to 40%-SUV_{MAX} mask as a conventional threshold based method.

Predicting ^{177}Lu-PSMA therapy response has been the second goal of this study. To this end, we calculate radiomics features based on the U-Net predicted masks. As radiomics analysis has been successfully used in many oncological studies for treatment response prediction and analysis of overall survival [5,14,29,30], combining automated segmentation with radiomics analysis for multimodal ^{68}Ga-PSMA-PET/CT findings is another contribution of our method. Results of the classification task confirm the potential of a fully automated approach, even though the comparison to predictions based on manual segmentation still indicates room for improvement. In the future, we plan to explore an end to end prediction of treatment response using a deep neural network. However, we expect that successfully training such an approach might require a larger cohort.

In this study, we focused on ^{68}Ga-PSMA-PET/CT scans and used a retrospective dataset from a single NM center using a single PET/CT scanner. However, the findings need to be further compared to that of other scanners as well as other biomarkers such as fluorodeoxyglucose (FDG). To improve these preliminary results, both the U-Net based segmentation as well as the radiomics analysis pipelines should be enhanced. Furthermore, to implement decision support tools which can take part in clinical routines in near future, we plan to include PET/CT images from different scanners and centers as well as other biomarkers.

5   Conclusion

Successful prediction of ^{177}Lu-PSMA treatment response would have a major impact on clinical decisions in patients with advanced prostate carcinoma. To our knowledge, we present the first fully automated system for this task. It is based on applying a multi-channel U-Net to multimodal ^{68}Ga-PSMA-PET/CT scans, which automatically delineates pathological uptake with a high accuracy. Supervised machine learning is then applied to radiomics features to predict treatment response. We expect that training data from larger studies will further increase the accuracy achieved by systems like ours, and will permit assessing the generalizability of the results.

Data and Code Availability. Due to German regulations on medical data availability, we cannot disclose the data, however all the data would be available for review on-site. The in-house developed code is available online at https://gitlab.com/Moazemi/pet-ct-u-net.

References

1. Ferlay, J., et al.: Global Cancer Observatory: Cancer Today. International Agency for Research on Cancer, Lyon. https://gco.iarc.fr/today. Accessed 30 June 2021
2. Jin, S., Li, D., Wang, H., Yin, Y.: Registration of PET and CT images based on multiresolution gradient of mutual information demons algorithm for positioning esophageal cancer patients. J. Appl. Clin. Med. Phys. **14**(1), 55–61 (2013). https://doi.org/10.1120/jacmp.v14i1.3931
3. Bundschuh, R.A., et al.: Textural parameters of tumor heterogeneity in 18F-FDG PET/CT for therapy response assessment and prognosis in patients with locally advanced rectal cancer. J. Nucl. Med. **55**(6), 891–897 (2014). https://doi.org/10.2967/jnumed.113.127340
4. Bang, J.-I., et al.: Prediction of neoadjuvant radiation chemotherapy response and survival using pretreatment [^{18}F]FDG PET/CT scans in locally advanced rectal cancer. Eur. J. Nucl. Med. Mol. Imaging **43**(3), 422–431 (2015). https://doi.org/10.1007/s00259-015-3180-9

5. Ypsilantis, P.P., et al.: Predicting response to neoadjuvant chemotherapy with PET imaging using convolutional neural networks. PLoS One **10**(9) (2015). https://doi.org/10.1371/journal.pone.0137036

6. Gorgi Zadeh, S., et al.: CNNs enable accurate and fast segmentation of Drusen in optical coherence tomography. In: Cardoso, M.J. (ed.) DLMIA/ML-CDS -2017. LNCS, vol. 10553, pp. 65–73. Springer, Cham (2017). https://doi.org/10.1007/978-3-319-67558-9_8

7. Selvaganesan, K., et al.: Robust, atlas-free, automatic segmentation of brain MRI in health and disease. Heliyon. **5**(2), e01226 (2019). https://doi.org/10.1016/j.heliyon.2019.e01226

8. Ronneberger, O., Fischer, P., Brox, T.: U-Net: convolutional networks for biomedical image segmentation. In: Navab, N., Hornegger, J., Wells, W.M., Frangi, A.F. (eds.) MICCAI 2015. LNCS, vol. 9351, pp. 234–241. Springer, Cham (2015). https://doi.org/10.1007/978-3-319-24574-4_28

9. Hatt, M., et al.: The first MICCAI challenge on PET tumor segmentation. Med. Image Ana. **44**, 177–195 (2018). https://doi.org/10.1016/j.media.2017.12.007. ISSN 1361–8415

10. Li, L., Zhao, X., Lu, W., Tan, S.: Deep learning for variational multimodality tumor segmentation in PET/CT. Neurocomputing **392**, 277–295 (2020). https://doi.org/10.1016/j.neucom.2018.10.099. ISSN 0925–2312

11. Liu, L., Cheng, J., Quan, Q., Wu, F.X., Wang, Y. P., Wang, J.: A survey on U-shaped networks in medical image segmentations. Neurocomputing **409**, 244–258 (2020). https://doi.org/10.1016/j.neucom.2020.05.070. ISSN 0925–2312

12. Beukinga, R.J., et al.: Predicting response to neoadjuvant chemoradiotherapy in esophageal cancer with textural features derived from pretreatment 18F-FDG PET/CT imaging. J. Nucl. Med. **58**(5), 723–729 (2017). https://doi.org/10.2967/jnumed.116.180299

13. Cheng, J.Z., et al.: Computer-aided diagnosis with deep learning architecture: applications to breast lesions in US images and pulmonary nodules in CT scans. Scientific reports, vol. 6 (2016). https://doi.org/10.1038/srep24454

14. Moazemi, S., et al.: Decision-support for treatment with 177Lu-PSMA: machine learning predicts response with high accuracy based on PSMA-PET/CT and clinical parameters. Ann. Transl. Med. 9,9, 818 (2021). https://doi.org/10.21037/atm-20-6446

15. Parisot, C.: The DICOM standard. Int. J. Cardiac. Imag. **11**, 171–177 (1995). https://doi.org/10.1007/BF01143137

16. Abadi, M., et al.: TensorFlow: large-scale machine learning on heterogeneous systems. tensorflow.org (2015). https://doi.org/10.5281/zenodo.4724125

17. Chollet, F.: Keras. GitHub repository (2015). https://github.com/fchollet/keras

18. Dice, L.R.: Measures of the amount of ecologic association between species. Ecology **26**(3), 297–302 (1945). https://doi.org/10.2307/1932409

19. van Griethuysen, J.J.M., et al.: Computational radiomics system to decode the radiographic phenotype. Cancer Res. **77**(21), e104–e107 (2017). https://doi.org/10.1158/0008-5472.CAN-17-0339

20. Guyon, I., Weston, J., Barnhill, S., Vapnik, V.: Gene selection for cancer classification using support vector machines. Mach. Learn. **46**, 389–422 (2002). https://doi.org/10.1023/A:1012487302797

21. Official Company Website for the InterView FUSION: Software. https://www.mediso.de/Interview-fusion.html. Accessed 30 June 2021

22. Tomar, N.K.: Polyp Segmentation using UNET in TensorFlow 2.0. https://idiotdeveloper.com/polyp-segmentation-using-unet-in-tensorflow-2/. Accessed 30 June 2021

23. Wright, R.E.: Logistic regression. In: Grimm, L.G., Yarnold, P.R.: (eds.) Reading and Understanding Multivariate Statistics, pp. 217–244. American Psychological Association (1995)

24. Hearst, M.A., Dumais, S.T., Osuna, E., Platt, J., Scholkopf, B.: Support vector machines. IEEE Intell. Syst. Appl. **13**(4), 18–28 (1998). https://doi.org/10.1109/5254.708428

25. Simm, J., de Abril, I., Sugiyama, M.: Tree-based ensemble multi-task learning method for classification and regression **97**(6) (2014). http://CRAN.R-project.org/package=extraTrees

26. Breiman, L.: Random forests. Mach. Learn. **45**, 5–32 (2001). https://doi.org/10.1023/A:1010933404324

27. Moazemi, S., et al.: Machine learning facilitates hotspot classification in PSMA-PET/CT with nuclear medicine specialist accuracy. Diagnostics (Basel, Switzerland) **10**(9), 622 (2020). https://doi.org/10.3390/diagnostics10090622

28. Erle, A., Moazemi, S., Lütje, S., Essler, M., Schultz, T., Bundschuh, R.A.: Evaluating a machine learning tool for the classification of pathological uptake in whole-body PSMA-PET-CT scans. Tomography **7**, 301–312 (2021). https://doi.org/10.3390/tomography7030027

29. Vallieres, M., et al.: Radiomics strategies for risk assessment of tumour failure in head-and-neck cancer. Sci. Rep. **7**(1), e10117 (2017). https://doi.org/10.1038/s41598-017-10371-5

30. Moazemi, S., Erle, A., Lütje, S., Gaertner, F.C., Essler, M., Bundschuh, R.A.: Estimating the potential of radiomics features and radiomics signature from pretherapeutic PSMA-PET-CT scans and clinical data for prediction of overall survival when treated with 177Lu-PSMA. Diagnostics **11**(2), 186 (2021). https://doi.org/10.3390/diagnostics11020186

A Federated Multigraph Integration Approach for Connectional Brain Template Learning

Hızır Can Bayram and Islem Rekik[✉] [ID]

BASIRA Lab, Faculty of Computer and Informatics, Istanbul Technical University,
Istanbul, Turkey
irekik@itu.edu.tr
http://basira-lab.com

Abstract. The connectional brain template (CBT) is a compact representation (i.e., a single connectivity matrix) *multi-view brain networks* of a given population. CBTs are especially very powerful tools in brain dysconnectivity diagnosis as well as holistic brain mapping if they are learned properly – i.e., occupy the center of the given population. Even though accessing large-scale datasets is much easier nowadays, it is still challenging to upload all these clinical datasets in a server altogether due to the data privacy and sensitivity. Federated learning, on the other hand, has opened a new era for machine learning algorithms where different computers are trained together via a distributed system. Each computer (i.e., a client) connected to a server, trains a model with its local dataset and sends its learnt model weights back to the server. Then, the server aggregates these weights thereby outputting global model weights encapsulating information drawn from different datasets in a *privacy-preserving* manner. Such a pipeline endows the global model with a generalizability power as it implicitly benefits from the diversity of the local datasets. In this work, we propose *the first federated connectional brain template learning* (Fed-CBT) framework to learn how to integrate multi-view brain connectomic datasets collected by different hospitals into a single representative connectivity map. First, we choose a random fraction of hospitals to train our global model. Next, all hospitals send their model weights to the server to aggregate them. We also introduce a weighting method for aggregating model weights to take full benefit from all hospitals. Our model to the best of our knowledge is the first and only federated pipeline to estimate connectional brain templates using graph neural networks. Our Fed-CBT code is available at https://github.com/basiralab/Fed-CBT.

Keywords: Connectional brain templates · Federated learning · Multigraph integration · Graph neural networks · Brain connectivity

1 Introduction

Technological advances and publicly open datasets in network neuroscience have made it possible to understand the function and structure of the human

© Springer Nature Switzerland AG 2021
T. Syeda-Mahmood et al. (Eds.): ML-CDS 2021, LNCS 13050, pp. 36–47, 2021.
https://doi.org/10.1007/978-3-030-89847-2_4

brain [1, 2]. Such advances and datasets, in particular, help us better differentiate a population of healthy from unhealthy subjects by mapping and comparing their brain connectivities. Connectional brain templates, CBTs, are great templates of multimodal brain networks in terms of how they integrate complementary information from multi-view or multi-modal connectivity datasets for a given population [3–5]. A CBT can be a powerful representation of a population of multi-view brain connectomes if it is well-centered (i.e., achieves the minimum distance to all connectomes in the population) and discriminative (i.e., discriminates between brain states such as healthy and disordered) [5]. Nevertheless there are various barriers to estimating such a representative and holistic CBT from connectomic datasets collected from different sources (e.g., hospitals or clinical facilities).

Given a population of subjects, estimating a CBT is a challenging task due to the complexity, heterogeneity and high-dimensionality of multi-view connectomic datasets [5] as well as the inter-individual variability across subjects and brain views (e.g., structural and functional connectivities) [4]. To estimate such a brain template, [3] proposed clustering-based multi-view network fusion. [3] fuses different views of every subject into one view and then clusters these views based on the similarity of their connectivity patterns. Finally, it averages those views and generates a CBT of the population. Despite its good results, its performance heavily depends on the number of clusters used during the fusion step. To eliminate this problem, [4] offered netNorm framework, which forms a tensor containing common cross-view features across subjects and then fuses distinct layers of this tensor into a CBT of the given population in a non-linear manner. In spite of its promising results, [4] has various limitations. First, it uses Euclidean distance as a dissimilarity metric to choose the most representative connections; however, this prevents the framework from capturing non-linear patterns. Second, netNorm consists of independent components which inhibits the training of the proposed framework in an end-to-end fashion. Another attempt for estimating CBT is [5]. Deep Graph Normalizer, DGN for short, presents the first graph neural network (GNN) to learn a CBT from a given multigraph population. DGN also introduces a novel loss, SNL, that optimizes the model to create a CBT that is well-centered and representative. However, SNL loss function does not evaluate how topological soundness of generated CBT is [6]. To alleviate this, [6] designed MGN-Net framework, which involves a novel loss function that forces the model to learn a topologically sound CBT while preserving the population distinctive traits.

However, all these methods share a prime challenge. To produce a highly representative CBT of a particular brain state, one needs to cover all the normative variations of that state – which is only possible when collecting from different global sources (e.g., hospitals or clinical facilities). However, clinical data sharing is a sensitive research topic with diverse ethical ramifications with regard to data privacy and protection protocols. Federated learning, on the other hand, is an emerging field with lots of promises such as data privacy, decentralized training, data variability and so on [7]. It is mainly rooted in the concept of training local models, called clients, with their own datasets and sending their updated weights to a server. This server then aggregates those weights in order to have global weights gathered from clients without really using their local private datasets.

Federated learning has various medical applications. [8] offered a federated named entity recognition (NER) pipeline since data diversity is significant yet hard to gather in NER due to privacy issues. In this way, the global model in the pipeline can boost the training. [9] proposed an algorithm to detect COVID-19 while leveraging datasets in their own medical institutions using federated learning. They designed a dynamic fusion-based federated approach in order to alleviate the cost of transferring model updates and data heterogeneity of default setting of federated learning. [10] presented a federated pipeline to take advantage of big data for better understanding brain disorders. [11] introduced a medical relation extraction model to detect and classify semantic relationships from biomedical text using federated learning. Their method also includes knowledge distillation in order to defuse communication bottleneck. [12] suggested a federated learning approach to solve the problem of multi-site functional magnetic resonance imaging (fMRI) classification combined with two domain adaptation methods. [13] offered a new federated learning approach other than [14] which is a channel-based update algorithm, more suitable for medical data – among others. **Although promising, none of these methods addressed the problem of connectomic data integration across a pool of hospitals** [15]. To fill this gap, we propose a federated multigraph integration framework, namely Fed-CBT, in which different hospitals estimate their CBTs based on their local datasets and share their model weights with a server so that the server aggregates them in a way that the global model can benefit from all hospitals' datasets. **This work is a primer, proposing the first *federated* GNN-based framework for multi-view graph fusion.**

2   Proposed Method

In this section, we give details about the components of the proposed method. Fed-CBT strategy consists of two main components: A) Training c fraction of hospitals and sending the models' weights to the server and B) Averaging the models' weights with designated strategy and updating hospitals in a federated manner. First, a DGN model [5] is initialized in the server and its weights are shared with hospitals that will be actively trained in the current round. Hospitals that are not trained in the current round do not update their weights. Thus, they just pass back their weights to the server. Other hospitals, however, are trained with their own datasets. We noticed that updating some fraction of hospitals accelerates the global model convergence. In each round, each hospital (client) randomly samples a subset from its local dataset for training and passes only one brain multigraph (i.e., subject connectivity tensor) to the DGN model composed of 3 layers of edge-conditioned convolution [16] in every batch. Several tensor operations (Fig. 1) are applied to the output feature vector and a CBT that represents the population of the dataset of a hospital is then generated. The learned CBT is evaluated against

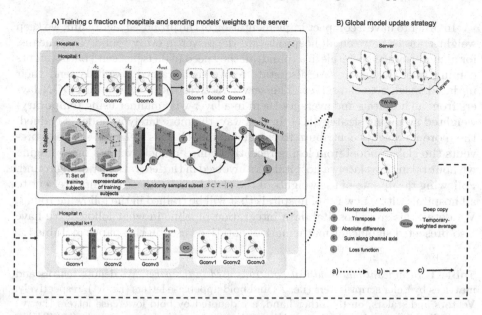

Fig. 1. *Federated connectional brain template (Fed-CBT) learning framework for integrating multi-view brain networks collected from different hospitals.* Here we illustrate one round of federated learning pipeline. Purple hospitals represent the hospitals trained during this round whereas blue ones represent those left-out. Arrow with point-like dash (a) implies that deep-copied weights of all layers are sent from hospitals to the server for averaging. Dashed arrow (b) implies that averaged weights of all layers are sent to all hospitals. Arrow (c), finally, implies the input-output flow between different operations. We give explanation of phase A and B next. **(A) Training c fraction of hospitals and sending models' weights to the server.** In each round, hospitals are randomly selected to be trained using their datasets. Features obtained from the last layer of DGN models are used to a generate CBT. After training, these hospitals send their updated model weights to the server while rest of the hospitals send what they receive. **(B) Global model update strategy.** Following the collection of the models' weights in the server, their weights are averaged with temporary-weighted averaging [17] in a layer-wise manner. Thus an updated version of the global model is generated. Next, weights of the updated global model are sent to hospitals trained in the current round. (Color figure online)

a randomly sampled subset of the local training dataset of the hospital using the loss function (Eq. 2) in order to optimize the model in a way that it generates a more centered CBT of the training dataset.

In order to have a compact implementation of the proposed algorithm, we keep weight transfer between all hospitals and the server in every round. This means, for all non-trained hospitals in a round, all they need to do is just send back weights coming from the server. We also state above that trained hospitals send their updated weights to the server. The server receives the weights of all DGN layers from all hospitals and averages them in a layer-wise manner using temporary-weighted averaging strategy [17]. In this way, the more lately a model is trained, the more its weights contribute to the global model. This weighted strategy prevents the global model from losing recently trained models weights by averaging without taking how lately a hospital gets involved in the federation training round. Following the update of the weights of the global model, these weights are sent to all hospitals with a new list of hospitals that will be trained in the following round. We detail this cross-hospital data integration pipeline in what follows. We have also presented significant mathematical notations used in this paper in Table 1.

Table 1. *Significant mathematical notations used in the paper.* We denote tensors and matrices by Euler script letters (i.e. $\mathcal{T}$) and bold uppercase letters (i.e. $\mathbf{M}$), respectively. Vectors and scalars, on the other hand, are denoted by bold lowercase letters (i.e. $\mathbf{v}$) and italic lowercase letters (i.e. s). We use italic uppercase letters for sets (i.e. TH).

Notation	Definition
c	Fraction of hospitals trained in each round
e	Number of local epochs in each round
n	Number of hospitals
TH^t	Set of hospitals trained in round t
H^t	Set of untrained hospitals in round t
T_j	Training set of multi-view brain networks of hospital j
n_v	Number of brain networks of a subject
n_r	Number of nodes (anatomical regions) in a brain network
$\mathcal{T}_{js}$	Tensor representation of subject s in hospital j, $\mathcal{T}_{js} \in \mathbb{R}^{n_r \times n_r \times n_v}$
d_0	Number of initial attributes for each node
$\mathbf{V}^0$	A node attributes identity matrix, $\mathbf{V}^0 \in \mathbb{R}^{n_r \times d_0}$
l	Number of graph convolution layers used in DGN architecture
$\mathbf{v}_j^l$	Output vector of DGN architecture with l layers in hospital j, $\mathbf{v}_j^l \in \mathbb{R}^{n_r \times 1}$
$\mathbf{C}_j$	Estimated CBT $\mathbf{C}_j$, $\mathbf{C}_j \in \mathbb{R}^{n_r \times n_r}$
S	Set of randomly selected training sample indices
$\mathbf{C}_{js}$	Generated CBT for subject s of hospital c, $\mathbf{C}_{js} \in \mathbb{R}^{n_r \times n_r}$
λ_v	Normalization term for brain graph connectivity weights of view v used in SNL loss function
$\mathcal{W}_j^t$	Model weight of all layers of hospital j in round t
ts^j	The round where hospital j is trained the latest

A-Training c Fraction of Hospitals and Sending Models' Weights to the Server. We propose a federated pipeline with n hospital where n/c of them are selected randomly with replacement for training in each round. We follow [5] to estimate a CBT of a given multigraph population. We denote hospitals that will be trained in round t as $TH^t = \{TH_1^t, TH_2^t, ..., TH_k^t\}$ and those left untrained as $H^t = \{H_1^t, H_2^t, ..., H_{n-k}^t\}$. We represent the training set of multi-view brain networks of hospital j as $T_j = \{T_{j1}^1, ..., T_{ji}^v, ..., T_{jn}^{n_v}\}$, where T_{ji}^v denotes the vth view of subject i in hospital j. Given a population of subjects for a hospital j, each subject s in this population is represented by a tensor $\mathcal{T}_s^j \in \mathbb{R}^{n_r \times n_r \times n_v}$. Here, n_v and n_r indicate the number of brain networks and number of nodes in these brain networks, respectively. Each node denotes an anatomical brain region of interest (ROI) and the weight of an edge connecting two regions models their relationship (e.g., morphological similarity or correlation in neural activity). Aside from edge multi-view attributes defined by a tensor $\mathcal{T}_s^j$, DGN model[1] initializes the node attributes as identity matrix $\mathbf{V}^0 \in \mathbb{R}^{n_r \times d_0}$, where d_0 denotes the number of initial attributes for each node. We used such initialization strategy since the brain connectomes we used in this study are not annotated (i.e., nodes have no extra attributes).

The local multi-view brain connectivity dataset of each hospital $j \in TH$ is fed into a DGN model consisting of 3 graph convolutional layers with edge-conditioned convolution operation [16], each followed by ReLU activation function. Let l denote the layer index in the DGN model. Following forward propagation, the output $\mathbf{V}_j^l = \left[\mathbf{v}_{j1}^l, \mathbf{v}_{j2}^l, ..., \mathbf{v}_{jn_r}^l\right]^T$ is obtained, denoting the output of the jth hospital and lth layer. $\mathbf{V}_j^l$ is copied across x-axis n_r times to obtain $\mathcal{R}_k \in \mathbb{R}^{n_r \times n_r \times d_L}$. $\mathcal{R}_k$ is also transposed to obtain $\mathcal{R}_k^T$. Finally, element-wise absolute difference of $\mathcal{R}_k$ and $\mathcal{R}_k^T$ is computed. We adopt such operation since our brain connectivity weights are derived from an absolute difference operation between brain region attributes. More details about our evaluation data are provided in the following section. The resulting tensor is summed along z-axis to estimate the final CBT $\mathbf{C}_j \in \mathbb{R}^{n_r \times n_r}$. The generated CBT is evaluated against a random subset of the training subjects with SNL loss function for regularization [5]. Given the generated CBT $\mathbf{C}_{js}$ for subject s of hospital j and a random subset S of training subjects indices, SNL loss function is defined as follows [5]:

$$SNL_{js} = \sum_{v=1}^{n_v} \sum_{i \in S} \left\| \mathbf{C}_{js} - \mathbf{T}_{ji}^v \right\|_F \times \lambda_v; \qquad \min_{\mathbf{W}_1, \mathbf{b}_1 ... \mathbf{W}_l, \mathbf{b}_l} \frac{1}{|T_j|} \sum_{s=1}^{|T_j|} SNL_{js}$$

The λ_v is defined as $\lambda_v = \frac{\max\{\mu_j\}_{j=1}^{n_v}}{\mu_v}$, where μ_v is the mean of brain graph connectivity weights of view v.

Each hospital, either trained or not, computes the Frobenius distance between its own local testing dataset and the generated CBT $\mathbf{C}_k$ using their latest model weights. Following the backward propagation phase of the trained hospitals, updated model weights and their corresponding loss values are sent to the server.

---

[1] https://github.com/basiralab/DGN.

As for the remaining hospitals, weights of the models received from server and their corresponding losses are simply passed back to the server without training.

B-Global Model Update Strategy. In this part, we detail the federated optimization step. In this step, all N hospitals in the pipeline send their model weights to the server at the end of round t. We can denote hospital j's loss in round t as follows:

$$f_j(\mathcal{W}_j^t) = \frac{1}{|\mathcal{T}_j|} \sum_{i=1}^{T_j} SNL(\mathcal{T}_{ji}; \mathcal{W}_j^t)$$

Using the function above, we can express the federated loss function of all hospitals in round t as follows [14]:

$$f(\mathcal{W}^t) = \sum_{j=1}^{K} f_j(\mathcal{W}_j^t)$$

We, on the other hand, formulate the federated model's weights in the following way:

$$\mathcal{W}^{t+1} = \frac{1}{N} \sum_{n=1}^{N} \mathcal{W}_n^t$$

where, $\mathcal{W}_n^t$ denotes the weights of all layers of the nth hospital in the tth round and $\mathcal{W}^{t+1}$ is the federated weights of all layers in the following $(t+1)$ round.

However, we note that some hospitals can be trained more than others since the selection of hospitals to train in the next round is fully stochastic. In this scenario, equally averaging all models' weights might dilute the weights of the models trained more than others (i.e., vanishing weights). To eliminate such a problem we adopt a temporally weighted aggregation strategy [17]. In this way, the more recently a hospital is selected to be trained, the stronger the contribution of its weights to the following round. Adding this trick to the model aggregation strategy is formalized by the following equation:

$$\mathcal{W}^{t+1} = \sum_{n=1}^{N} \frac{\left(\frac{\exp}{2}\right)^{-(t-ts^n)}}{\sum_{n=1}^{N}\left(\frac{\exp}{2}\right)^{-(t-ts^n)}} \mathcal{W}_n^t$$

where t and ts^j denote the current round and the round where hospital j is trained the latest, respectively.

3   Results and Discussion

Dataset. We evaluated our Fed-CBT, federated DGN with temporary-weighted averaging (TW), and Fed-CBT w/o TW strategies against DGN model [5]

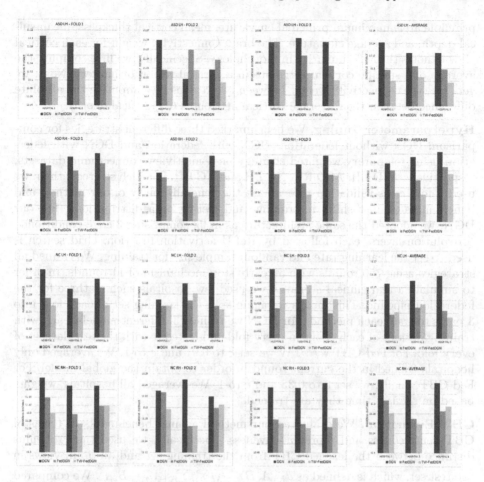

Fig. 2. *Centeredness comparison between CBTs generated by DGN, Fed-CBT w/o TW and Fed-CBT.* Different strategies are evaluated using the Frobenius distance between the learned CBT and testing brain networks. We kept testing sets of each fold of each dataset the same across different strategies. We also included the average Frobenius distances across all folds for each dataset. ASD: autism spectrum disorder. NC: normal control. LH: left hemisphere. RH: right hemisphere.

and against each other using normal control (NC) and autism spectrum disorder (ASD) dataset, collected from the Autism Brain Imaging Data Exchange ABIDE-I public dataset [18]. NC/ASD dataset consists of 186 subjects with normal control and 155 subjects with autism spectrum disorder involving both hemisphere, right and left. Cortical hemispheres are reconstructed from T1-weighted MRI using FreeSurfer [19]. Next, each one is they are parcellated into 35 ROIs using Desikan-Killiany Atlas [20]. For each cortical hemisphere, each individual in the dataset is represented by a connectivity tensor including 6 cortical morphological brain networks as introduced in [21–23]: cortical surface area, minimum principle area, maximum principal curvature, mean cortical thickness, mean sulcal depth, and average curvature. The brain connectivity weight for each cortical view denotes their dissimilarity in morphology. Specifically, we first compute the average of a specific cortical attribute in a cortical region of interest. Next, we generate the connectivity weight between two regions by computing the absolute difference between their average cortical attributes (e.g., sulcal depth).

Hyperparameter Tuning. We benchmarked three different strategies for comparison: DGN without federation, DGN with federation and DGN with federation using temporary-weighted strategy on 4 multiview-connectomic datasets, respectively: ASD LH, ASD RH, NC LH and NC RH. For each strategy, then, we used 3-fold cross-validation to evaluate the generalizability of our method and benchmarks against shifts in training and testing data distributions. For our DGN, we adopted the same architecture as in [5] comprising 3 edge-conditioned convolution layers, each followed by ReLU activation function. Grid search is used to set the learning rate and random sample size for training. We trained all strategies using 750 rounds with an early stop mechanism of 11 rounds. In order to simulate a real clinical scenario, we used few hospitals such as three for the federation pipeline. This means we again split the training set of each fold into 3 parts to train each method. For DGN we trained 3 different models, each with training dataset of each client in every fold and tested it with the test dataset of every fold. For Fed-CBT w/o TW we set c to 0.91 and e to 1. We averaged only hospitals trained in the current round in order to obtain the global model. For Fed-CBT, finally, we set c to 0.33 and e to 1. We averaged all hospitals' weights based on the last time they are trained.

CBT Evaluation. We evaluated our method against benchmarks in terms of CBT centeredness and representativeness. Specifically, we used the Frobenius distance between the learned CBT from the training set and all brain views in the test set, which is defined as $d_F(A, B) = \sqrt{\sum_i \sum_j |A_{ij} - B_{ij}|^2}$. We compared each hospital's Frobenius distance across different strategies for each fold of each dataset (Fig. 2). We also report the average Frobenius distance across folds for each dataset and each hospital. We found out that Fed-CBT and Fed-CBT w/o TW outperform DGN across all four datasets. We also note that Fed-CBT further outperforms Fed-CBT w/o TW in ASD LH, ASD RH and NC RH datasets. Figure 3 displays the estimated CBTs of ASD LH dataset for each strategy: DGN, Fed-CBT w/o TW and Fed-CBT.

Limitations and Future Directions. One key limitation of this work is that proposed federated multigraph integration approach has only been evaluated in the case where hospitals share the same number of brain connectivity views. We will extend our work by designing a novel federated integration strategy that handles variation in the number of the brain connectivity views. The communication cost between the server and clients can also be decreased by adopting efficient strategies as proposed in [24] and [25].

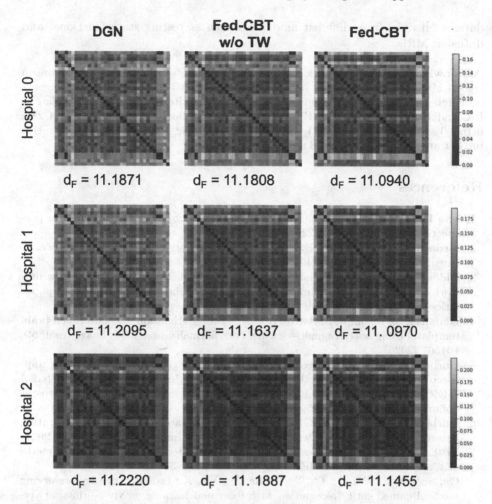

Fig. 3. *The learned CBTs based on different strategies.* ASD RH dataset is partitioned into three folds where each fold is further partitioned into three sets, each assigned to a single hospital. Here we display the learned CBT using fold 1 by DGN, Fed-CBT w/o TW and Fed-CBT strategies, respectively.

4   Conclusion

In this work, we presented the first *federated brain multigraph integration* framework across a set of hospitals where multi-view brain connectivity datasets are locally collected. We showed that hospitals can share their knowledge and boost the representativeness of their locally learned CBTs without resorting to any clinical data sharing. In our future work, we will extend our proposed Fed-CBT to handle connectomic datasets coming in varying number of views across hospitals. We will also evaluate our framework on multi-modal brain connectivity datasets derived from different modalities such as resting state functional and diffusion MRIs.

Acknowledgments. This work was funded by generous grants from the European H2020 Marie Sklodowska-Curie action (grant no. 101003403, http://basira-lab.com/normnets/) to I.R. and the Scientific and Technological Research Council of Turkey to I.R. under the TUBITAK 2232 Fellowship for Outstanding Researchers (no. 118C288, http://basira-lab.com/reprime/). However, all scientific contributions made in this project are owned and approved solely by the authors.

References

1. Van Essen, D.C., et al.: The human connectome project: a data acquisition perspective. Neuroimage **62**, 2222–2231 (2012)
2. Medaglia, J.D., Lynall, M.E., Bassett, D.S.: Cognitive network neuroscience. J. Cogn. Neurosci. **27**, 1471–1491 (2015)
3. Dhifallah, S., Rekik, I., Initiative, A.D.N., et al.: Clustering-based multi-view network fusion for estimating brain network atlases of healthy and disordered populations. J. Neurosci. Methods **311**, 426–435 (2019)
4. Dhifallah, S., Rekik, I., Initiative, A.D.N., et al.: Estimation of connectional brain templates using selective multi-view network normalization. Med. Image Anal. **59**, 101567 (2020)
5. Gurbuz, M.B., Rekik, I.: Deep graph normalizer: a geometric deep learning approach for estimating connectional brain templates. In: Martel, A.L., et al. (eds.) MICCAI 2020. LNCS, vol. 12267, pp. 155–165. Springer, Cham (2020). https://doi.org/10.1007/978-3-030-59728-3_16
6. Gürbüz, M.B., Rekik, I.: MGN-Net: a multi-view graph normalizer for integrating heterogeneous biological network populations. Med. Image Anal. **71**, 102059 (2021)
7. Li, T., Sahu, A.K., Talwalkar, A., Smith, V.: Federated learning: challenges, methods, and future directions. IEEE Sig. Process. Mag. **37**, 50–60 (2020)
8. Ge, S., Wu, F., Wu, C., Qi, T., Huang, Y., Xie, X.: FedNER: privacy-preserving medical named entity recognition with federated learning. arXiv e-prints. arXiv-2003 (2020)
9. Zhang, W., et al.: Dynamic fusion-based federated learning for COVID-19 detection. IEEE Internet Things J. (2021)
10. Silva, S., Gutman, B.A., Romero, E., Thompson, P.M., Altmann, A., Lorenzi, M.: Federated learning in distributed medical databases: meta-analysis of large-scale subcortical brain data. In: 2019 IEEE 16th International Symposium on Biomedical Imaging (ISBI 2019), pp. 270–274. IEEE (2019)
11. Sui, D., Chen, Y., Zhao, J., Jia, Y., Xie, Y., Sun, W.: FedED: federated learning via ensemble distillation for medical relation extraction. In: Proceedings of the 2020 Conference on Empirical Methods in Natural Language Processing (EMNLP), pp. 2118–2128 (2020)
12. Li, X., Gu, Y., Dvornek, N., Staib, L.H., Ventola, P., Duncan, J.S.: Multi-site fMRI analysis using privacy-preserving federated learning and domain adaptation: abide results. Med. Image Anal. **65**, 101765 (2020)
13. Shao, R., He, H., Liu, H., Liu, D.: Stochastic channel-based federated learning for medical data privacy preserving. arXiv preprint arXiv:1910.11160 (2019)

14. McMahan, B., Moore, E., Ramage, D., Hampson, S., Aguera y Arcas, B.: Communication-efficient learning of deep networks from decentralized data. In: Artificial Intelligence and Statistics. PMLR, pp. 1273–1282 (2017)
15. Bessadok, A., Mahjoub, M.A., Rekik, I.: Graph neural networks in network neuroscience. arXiv preprint arXiv:2106.03535 (2021)
16. Simonovsky, M., Komodakis, N.: Dynamic edge-conditioned filters in convolutional neural networks on graphs, pp. 29–38 (2017)
17. Chen, Y., Sun, X., Jin, Y.: Communication-efficient federated deep learning with layerwise asynchronous model update and temporally weighted aggregation. IEEE Trans. Neural Netw. Learn. Syst. 31, 4229–4238 (2019)
18. Di Martino, A., et al.: The autism brain imaging data exchange: towards a large-scale evaluation of the intrinsic brain architecture in autism. Mol. Psychiatry 19, 659–667 (2014)
19. Fischl, B.: FreeSurfer. Neuroimage 62, 774–781 (2012)
20. Desikan, R.S., et al.: An automated labeling system for subdividing the human cerebral cortex on MRI scans into gyral based regions of interest. Neuroimage 31, 968–980 (2006)
21. Mahjoub, I., Mahjoub, M.A., Rekik, I.: Brain multiplexes reveal morphological connectional biomarkers fingerprinting late brain dementia states. Sci. Rep. 8, 1–14 (2018)
22. Corps, J., Rekik, I.: Morphological brain age prediction using multi-view brain networks derived from cortical morphology in healthy and disordered participants. Sci. Rep. 9, 1–10 (2019)
23. Bilgen, I., Guvercin, G., Rekik, I.: Machine learning methods for brain network classification: application to autism diagnosis using cortical morphological networks. J. Neurosci. Methods 343, 108799 (2020)
24. Muhammad, K., et al.: FedFast: going beyond average for faster training of federated recommender systems. In: Proceedings of the 26th ACM SIGKDD International Conference on Knowledge Discovery & Data Mining, pp. 1234–1242 (2020)
25. Sun, Y., Zhou, S., Gündüz, D.: Energy-aware analog aggregation for federated learning with redundant data. In: ICC 2020–2020 IEEE International Conference on Communications (ICC), pp. 1–7. IEEE (2020)

SAMA: Spatially-Aware Multimodal Network with Attention For Early Lung Cancer Diagnosis

Mafe Roa[1,2(✉)] [iD], Laura Daza[1,2] [iD], Maria Escobar[1,2] [iD], Angela Castillo[1,2] [iD], and Pablo Arbelaez[1,2] [iD]

[1] Center for Research and Formation in Artificial Intelligence, Universidad de Los Andes, Bogotá, Colombia
{mf.roa,la.daza10,mc.escobar11,a.castillo13,pa.arbelaez}@uniandes.edu.co
[2] Department of Biomedical Engineering, Universidad de los Andes, Bogotá 111711, Colombia

Abstract. Lung cancer is the deadliest cancer worldwide. This fact has led to increased development of medical and computational methods to improve early diagnosis, aiming at reducing its fatality rate. Radiologists conduct lung cancer screening and diagnosis by localizing and characterizing pathologies. Therefore, there is an inherent relationship between visual clinical findings and spatial location in the images. However, in previous work, this spatial relationship between multimodal data has not been exploited. In this work, we propose a Spatially-Aware Multimodal Network with Attention (SAMA) for early lung cancer diagnosis. Our approach takes advantage of the spatial relationship between visual and clinical information, emulating the diagnostic process of the specialist. Specifically, we propose a multimodal fusion module composed of dynamic filtering of visual features with clinical data followed by a channel attention mechanism. We provide empirical evidence of the potential of SAMA to integrate spatially visual and clinical information. Our method outperforms by 14.3% the state-of-the-art method in the LUng CAncer Screening with Multimodal Biomarkers Dataset.

Keywords: Multimodal data · Lung cancer · Dynamic filters · Computer aided diagnosis

1 Introduction

Lung cancer is responsible for most cancer-related deaths worldwide [17]. Survival rates are highly dependent on early diagnosis. Nonetheless, this is a difficult task that requires highly specialized physicians and is prone to errors due to the

Electronic supplementary material The online version of this chapter (https://doi.org/10.1007/978-3-030-89847-2_5) contains supplementary material, which is available to authorized users.

© Springer Nature Switzerland AG 2021
T. Syeda-Mahmood et al. (Eds.): ML-CDS 2021, LNCS 13050, pp. 48–58, 2021.
https://doi.org/10.1007/978-3-030-89847-2_5

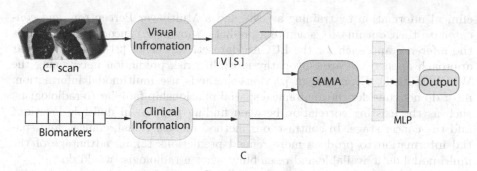

Fig. 1. SAMA exploits multimodal data by spatially combining visual and clinical information. We process visual and clinical information in two parallel branches. Their outputs go through the SAMA module, which fuses information using spatially-aware and channel attention mechanisms. The final representation passes through a MLP to generate a single prediction.

hard-to-see nodules, from which nearly 35% are missed in initial screenings [18]. Consequently, few patients are diagnosed on time with this type of cancer.

Early lung cancer diagnosis benefits from the increasing development of Computer-Aided Diagnosis (CAD) tools for tasks such as lung nodule detection, segmentation, and prognosis. In particular, Astaraki *et al.* [2] propose an autoencoder that removes the abnormalities in CT scans improving detection and segmentation. Zhang *et al.* [24] propose a dataset and a method that focuses on solid lung nodules to address uncertainty in the detection task. Besides, Moriya *et al.* present an unsupervised method for lung nodule segmentation in micro-CT scans [14]. Also, Huang *et al.* [7], and Ardila *et al.* [1] use previous CT scans for the same patient to predict the cancer incidence; and Li *et al.* [10] forecast the growth of the lung nodules in a time interval. Furthermore, Shaffie *et al.* [16] integrate visual information and bio-markers obtained from a single exhaled breath to classify lung nodules. All of these previous works aim to mimic the radiologists' diagnostic process and to automate it. Nonetheless, most lung cancer CAD methods consider only visual information, notwithstanding that physicians use imaging tests in conjunction with clinical history and patients' characteristics to support diagnosis [4,8]. The recently proposed LUng CAncer Screening with Multimodal Biomarkers (LUCAS) [3] dataset is the first experimental framework for early lung cancer diagnosis that provides clinical information considered by physicians in addition to chest CT scans. Thus it is a promising experimental framework to improve lung cancer CAD, and we use it as our experimental test bed.

Exploiting multimodal information is crucial, and CAD methods in medical tasks using this kind of data have shown promising results. For instance, Lara *et al.* [9] learn the alignment of visual and language information to perform histological images' classification. Furthermore, Tam *et al.* [19] propose a weakly supervised architecture for detection and visual grounding for pneumonia and pneumothorax studies. Yu *et al.* [23] use facial motion weaknesses and speech inability for rapid stroke diagnosis. Daza *et al.* [3] process visual and clinical information by training a CNN and a Multilayer Perceptron, and concatenate their outputs to generate a prediction for lung cancer, this method is the reference approach for the LUCAS dataset. Xia *et al.* [21] follow a similar approach applied to cardiomyopathy mortality risk prediction maximizing the WMW statistic. Even though previous methods use multimodal information, most do not consider the anatomical spatial relationship familiar to radiologists, such as the existing correlation between finding nodules in multiple locations and the cancer stage. In contrast, our method merges visual, clinical, and spatial information to produce more refined predictions taking advantage of the multimodal data available and resembling what a radiologist would do.

In this paper we present a **S**patially-**A**ware **M**ultimodal network with **A**ttention (SAMA) for early lung cancer diagnosis. SAMA is a novel method for multimodal analysis of lung cancer that comprises two parallel branches trained with visual and clinical information; and a SAMA module that combines their output. Figure 1 shows an overview of our method. Our SAMA module has two sequential components. First, the spatially-aware attention component performs dynamic filtering [11] of visual features with clinical data. Then the fusion with channel attention component performs a channel attention mechanism and fuses the multimodal information. Our main contribution is the formulation of multimodal fusion of visual and clinical information as a grounding task, which allows us to build on the idea of dynamic filters [11] as fusion strategy.

We empirically validate the relevance of each of SAMA's components and show evidence of the relation between the dynamic filtering response and the specialists' reports. Furthermore, we obtain an absolute improvement of 14% in F1-score over the state-of-the-art. To ensure reproducibility, our code can be found in https://github.com/BCV-Uniandes/SAMA.

2   Method

We propose SAMA, an automated method that predicts the probability of a patient having lung cancer using multimodal information (i.e., visual and clinical information). With our method, we emulate the screening process followed by radiologists when evaluating a medical image. Figure 1 illustrates our proposed multimodal architecture. First, our method processes information in two parallel branches, one for each modality of data available. For the visual information, we use the backbone proposed in [3] outputting a visual representation V. Likewise, the clinical information goes through an MLP outputting a clinical representation C. We process the two branches' output using our SAMA module (Fig. 2), which performs spatially-aware attention and then fusion with channel attention. Lastly, the fused representation obtained goes through an MLP to produce the final prediction.

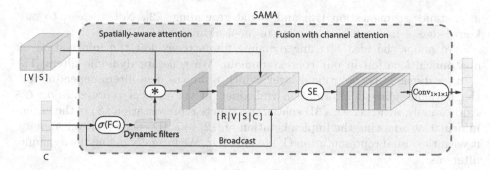

Fig. 2. SAMA module comprises two sequential components: *i)* spatially-aware attention;and *ii)* fusion with channel attention. Dynamic filters are generated from clinical representation (*C*) and convolved with visual (*V*) and 3D spatial coordinate (*S*) representations. Filters' response (*R*) is concatenated with visual, spatial and clinical information. The final multimodal representation goes through an Squeeze-and-Excitation block and a pointwise convolution to fuse the multimodal information.

2.1 SAMA Module

Spatially-Aware Attention. Handling information from diverse sources (i.e., multimodal information) is closer to the way humans process information and learn from it. For instance, in CT scans screening, visual information and patients' clinical history are correlated to an anatomical location in the volume. Radiologists consider this location to find visual characteristics related to specific pathologies. In lung cancer diagnosis, identifying the nodules' location is crucial for characterizing them and defining the patients' cancer stage.

In a non-clinical context, visual and language information is related to object location in video tracking. Li *et al.* proposed a novel in-video object tracking method based on language specification [11], which generates dynamic filters according to a text query. The method proposed by Li *et al.* has three different variants: use of only language, only images, or a join or multimodal approach. In the lingual specification variant, they proposed a dynamic convolutional layer. This layer generates new convolutional filters (dynamic filters) *f* in real-time according to a text query. Therefore, dynamic filters are target-specific visual filters based on the language input of the network. Dynamic convolutional layers are also employed in the multimodal variant of Li *et al.* method, where the filters are generated using a CNN that generates the visual features of the target. When they have obtained the dynamic filters using language information, they convolve them with the image (*I*), obtaining the response map *R* of the image to the filters. These maps are formally defined as:

$$R = f * I \tag{1}$$

Dynamic filters can be thought of as specialized and fine-tuned according to the semantics in the language specification. This approach, has been transferred to different natural image tasks involving linguistic and visual information, such as instance segmentation [13] and visual grounding [22]. Nonetheless, to our knowledge, it had not been applied to medical images until now.

We adopt the idea of using dynamic filters to exploit the information of anatomical location in our context domain. We generate dynamic filters [11] using a dynamic convolutional layer, which produces new filters depending on the patients' clinical information embedded in the clinical representation C. Additionally, we generate a 3D spatial coordinate representation (S) of the visual information adapting the implementation of [12] to a 3D space and concatenate it with the visual representation (V) to enrich it. We formally define the dynamic filters as:

$$f_k = \sigma(FC), \quad k = 1, ..., K. \tag{2}$$

Where σ is the sigmoid function, FC is a fully connected layer, and K is a hyperparameter defining the number of dynamic filters generated. Each filter f_k has the same number of channels as the sum of the channels of V and S. We calculate a dynamic response map R as the convolution of the concatenatation of V and S with the dynamic filters f_k:

$$R = \{f_k * [V|S]\}_{k=1}^{K}. \tag{3}$$

Fusion with Channel Attention. We concatenate the response map R with the visual representation V, the 3D spatial coordinate representation S, and the clinical representation C broadcasted to match each spatial location, all along the channel dimension.

We pass this concatenation through a Squeeze-and-Excitation block (SE) [5], a channel attention mechanism that weights channels according to their relevance. Finally, this weighted representation goes through a pointwise convolutional layer to fuse all the multimodal information at each location independently, producing the fused representation O:

$$O = \text{Conv}_{1 \times 1 \times 1}(\text{SE}([R|V|S|C])). \tag{4}$$

Finally, the output O goes through a Multilayer perceptron (MLP) with one hidden layer. The network outputs a predicted score representing the probability of having cancer.

3 Experimental Setup

3.1 Dataset

We conduct experiments in the LUng CAncer Screening with Multimodal Biomarkers (LUCAS) [3] dataset, a realistic experimental framework containing 830 CT scans with 72 positive samples. Each sample has corresponding anonymized clinical data from a radiologist's report, including variables denoting approximate anatomical location (e.g., superior, inferior, left, or right lobes), nodule size, among others. We evaluate the performance using the F1-score and

Table 1. Comparison with the state-of-the-art method LUCAS [3].

Model	F1	AP
LUCAS [3]	0.198 (0.042)	0.091 (0.019)
SAMA	**0.341 (0.058)**	**0.251 (0.061)**

average precision score (AP) metrics, where the higher the value, the better performance. The maximum F1-score [6] is defined as the weighted average of the precision and recall. Acknowledging the high imbalance in the dataset we use 5-fold cross-validation. We retrain the LUCAS baseline model with this experimental framework for a fair comparison.

3.2 Implementation Details

The visual branch backbone is the LUCAS [3] baseline architecture with 512 channels in the last layer. We train for 15 epochs and a batch size of 6. We use Adam optimizer with a learning rate of 1×10^{-3}, and a weight decay of 1×10^{-5}. We adopt Instance Normalization [20] and ReLU activation function. We implement an MLP for the clinical branch with one hidden layer followed by a ReLU activation function with a final output size of 256. For the SAMA module, we define $K = 10$ dynamic filters. We train our model on a 12 GB Titan XP GPU for 15 h and total estimated carbon emission of 1.62 kg CO_2 eq.

4 Results

Table 1 compares our method's performance with the state-of-the-art method LUCAS [3] and shows the average of the 5-fold metrics; standard deviation appears in parenthesis. We show results for each fold in the Supplementary Material. We outperform the state-of-the-art results in F1-score by 14.3% and 16% in AP, on average. The metrics' improvement shows the advantage of using our multimodal fusion approach over an MLP. SAMA exploits the different types of data available and the interaction and relation inherent to the multimodal data.

4.1 Control Experiments

Multimodal Data. Table 2 shows the experiments performed to evaluate the complementary of clinical and visual information. Here we evidence that using multimodal data with our SAMA module, the performance is improved for both metrics. The results for multimodal data show that the visual and clinical data have a complementary nature, and training with both improves the performance compared to the training with any of them alone.

Table 2. Experiments to validate the complementary nature of multimodal data.

Model	F1	AP
Visual branch	0.191 (0.023)	0.103 (0.029)
Clinical branch	0.292 (0.064)	0.239 (0.071)
SAMA (multimodal)	**0.341 (0.058)**	**0.251 (0.061)**

Table 3. Ablation experiments to assess the effect of each component of our method.

Dynamic filters	SE block	F1	AP
-	✓	0.273 (0.061)	0.206 (0.044)
✓	-	0.265 (0.049)	0.174 (0.042)
✓	✓	**0.341 (0.058)**	**0.251 (0.061)**

General Architecture. Table 3 shows the effects of training SAMA by removing SAMA's components. Removing the dynamic filters and the spatial location information shows a reduction in performance by 6.8% in F1 and 4.5% in AP Here, the dynamic filters function as a spatially-aware attention mechanism relating visual information with approximate anatomical locations in the clinical data. In the second row, removing the Squeeze-and-Excitation (SE) block causes a reduction in the performance by 7.6% in F1 and 7.7% in AP compared to SAMA. The SE block allows our model to focus on the most relevant information. The performance reduction produced by removing any of the components is comparable, showing that both the spatially-aware attention mechanism (dynamic filters) and the channel attention mechanism (SE block) are crucial for the overall model performance and complement each other.

3D Spatial Coordinate Representation and Dynamic Filters. Table 4 shows results where we compare our proposed method with a method trained without the dynamic filters but keeping the 3D spatial coordinate representation S component (first row) and trained without S but keeping the dynamic filters (second row). The former reduces SAMA's performance by 5.4% in F1 and 3.1% in AP. The latter decreases our method's performance by 5.3% in F1 and 4.7% in AP. These results show that the dynamic filters and 3D spatial coordinate representation are both essential components of SAMA. The ensemble of these components associates the clinical information to a specific location in the volume, supporting the spatially-aware attention mechanism.

Dynamic Filters. Table 5 shows results for experiments with $K = \{4, 8, 10, 16, 32\}$ to evaluate the effect of having a different number of dynamic filters according to performance. Results show that the best performance is obtained with 16 filters.

Table 4. Ablation experiments to validate the importance of the 3D spatial coordinate representation (S) component.

Dynamic filters	S	F1	AP
-	✓	0.287 (0.053)	0.22 (0.072)
✓	-	0.288 (0.057)	0.204 (0.065)
✓	✓	**0.341 (0.058)**	**0.251 (0.061)**

Table 5. Experiments on different numbers of dynamic filters K.

K	F1	AP
4	0.288 (0.077)	0.202 (0.048)
8	0.287 (0.043)	0.166 (0.034)
10	0.265 (0.06)	0.163 (0.043)
16	**0.341 (0.058)**	**0.251 (0.061)**
32	0.305 (0.075)	0.224 (0.087)

Figure 3 shows qualitative results to empirically illustrate the dynamic filters' spatially-aware attention. We take each dynamic filter response map R_k, upsample, and overlay it to the original CT scan to inspect our model's spatially-aware attention. We show examples of the response maps for seven different patients. By comparing the spatially-aware attention locations with the clinical information the radiologists identified, they correspond to either the nodule's or other pathology's approximate anatomical location. For instance, in the second column, the patient's clinical report described two different lung nodules, and SAMA focuses on their approximate location. Furthermore, first column shows the results for a patient that was diagnosed with tree-in-bud pathology, small centrilobular nodules of soft-tissue attenuation connected to multiple branching linear structures of similar caliber that originate from a single stalk [15]. This pattern is easily recognizable as it resembles a budding tree or the objects used in the game of jacks. SAMA identifies the location of the pathology relating to the information given by the radiologist and the visual information. We show more detailed examples of qualitative results in the Supplementary Material. These results show an unsupervised intermediate representation that emphasizes our method's central idea: employing multimodal information to spatially express the clinical information given by the radiologist, enriching the final representation, and producing better predictions.

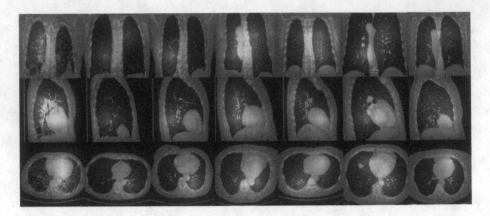

Fig. 3. Qualitative results of the dynamic filter response. SAMA focuses attention where radiologist identified a pathology. Results for seven different patients shown in columns. Coronal, Sagital and Axial planes in the rows.

5   Conclusions

We present SAMA, a method for exploiting multimodal information by spatially relating visual and clinical data for early lung cancer diagnosis using the LUCAS [3] dataset and outperform the state-of-the-art. Our model processes multimodal information in two parallel branches and merges these representations in a SAMA module where spatial information is related to the visual and clinical information using a spatially-aware attention mechanism and a 3D spatial coordinate representation. Furthermore, we use a channel attention mechanism that gives a higher weight to the channels that give more relevant information to improve the final prediction. SAMA merges spatially-aware and channel attention mechanisms to resemble the radiologists' diagnostic process. The multimodal fusion strategy for visual and clinical data we propose could be useful in other clinical contexts to improve multimodal tasks.

Acknowledgments. We thank AstraZeneca Colombia and the Lung Ambition Alliance for a research grant used for the development of this project.

References

1. Ardila, D., et al.: End-to-end lung cancer screening with three-dimensional deep learning on low-dose chest computed tomography. Nat. Med. **25**(6), 954–961 (2019)
2. Astaraki, M., Toma-Dasu, I., Smedby, Ö., Wang, C.: Normal appearance autoencoder for lung cancer detection and segmentation. In: Shen, D., Liu, T., Peters, T.M., Staib, L.H., Essert, C., Zhou, S., Yap, P.-T., Khan, A. (eds.) MICCAI 2019. LNCS, vol. 11769, pp. 249–256. Springer, Cham (2019). https://doi.org/10.1007/978-3-030-32226-7_28

3. Daza, L., Castillo, A., Escobar, M., Valencia, S., Pinzón, B., Arbeláez, P.: LUCAS: lung cancer screening with multimodal biomarkers. In: Syeda-Mahmood, T., Drechsler, K., Greenspan, H., Madabhushi, A., Karargyris, A., Linguraru, M.G., Oyarzun Laura, C., Shekhar, R., Wesarg, S., González Ballester, M.Á., Erdt, M. (eds.) CLIP/ML-CDS -2020. LNCS, vol. 12445, pp. 115–124. Springer, Cham (2020). https://doi.org/10.1007/978-3-030-60946-7_12
4. Gould, M.K., et al.: Evaluation of individuals with pulmonary nodules: when is it lung cancer?: diagnosis and management of lung cancer: American college of chest physicians evidence-based clinical practice guidelines. Chest **143**(5), e93S-e120S (2013)
5. Hu, J., Shen, L., Sun, G.: Squeeze-and-excitation networks. In: Proceedings of the IEEE conference on computer vision and pattern recognition, pp. 7132–7141 (2018)
6. Huang, H., Xu, H., Wang, X., Silamu, W.: Maximum f1-score discriminative training criterion for automatic mispronunciation detection. IEEE/ACM Trans. Audio, Speech, Lang. Process. **23**(4), 787–797 (2015)
7. Huang, P., et al.: Prediction of lung cancer risk at follow-up screening with low-dose CT: a training and validation study of a deep learning method. Lancet Digit. Health **1**(7), e353–e362 (2019)
8. de Koning, H.J., et al.: Reduced lung-cancer mortality with volume CT screening in a randomized trial. N. Engl. J. Med. **382**(6), 503–513 (2020)
9. Lara, J.S., Contreras O., V.H., Otálora, S., Müller, H., González, F.A.: Multimodal latent semantic alignment for automated prostate tissue classification and retrieval. In: Martel, A.L., Abolmaesumi, P., Stoyanov, D., Mateus, D., Zuluaga, M.A., Zhou, S.K., Racoceanu, D., Joskowicz, L. (eds.) MICCAI 2020. LNCS, vol. 12265, pp. 572–581. Springer, Cham (2020). https://doi.org/10.1007/978-3-030-59722-1_55
10. Li, Y., et al.: Learning tumor growth via follow-up volume prediction for lung nodules. In: Martel, A.L., Abolmaesumi, P., Stoyanov, D., Mateus, D., Zuluaga, M.A., Zhou, S.K., Racoceanu, D., Joskowicz, L. (eds.) MICCAI 2020. LNCS, vol. 12266, pp. 508–517. Springer, Cham (2020). https://doi.org/10.1007/978-3-030-59725-2_49
11. Li, Z., Tao, R., Gavves, E., Snoek, C.G., Smeulders, A.W.: Tracking by natural language specification. In: Proceedings of the IEEE Conference on Computer Vision and Pattern Recognition, pp. 6495–6503 (2017)
12. Liu, C., Lin, Z., Shen, X., Yang, J., Lu, X., Yuille, A.: Recurrent multimodal interaction for referring image segmentation. In: Proceedings of the IEEE International Conference on Computer Vision, pp. 1271–1280 (2017)
13. Margffoy-Tuay, E., Pérez, J.C., Botero, E., Arbeláez, P.: Dynamic multimodal instance segmentation guided by natural language queries. In: Proceedings of the European Conference on Computer Vision (ECCV), pp. 630–645 (2018)
14. Moriya, T., et al.: Unsupervised segmentation of Micro-CT images of lung cancer specimen using deep generative models. In: Shen, D., Liu, T., Peters, T.M., Staib, L.H., Essert, C., Zhou, S., Yap, P.-T., Khan, A. (eds.) MICCAI 2019. LNCS, vol. 11769, pp. 240–248. Springer, Cham (2019). https://doi.org/10.1007/978-3-030-32226-7_27
15. Rossi, S.E., Franquet, T., Volpacchio, M., Giménez, A., Aguilar, G.: Tree-in-bud pattern at thin-section CT of the lungs: radiologic-pathologic overview. Radiographics **25**(3), 789–801 (2005)
16. Shaffie, A., et al.: A novel technology to integrate imaging and clinical markers for non-invasive diagnosis of lung cancer. Sci. Rep. **11**(1), 1–10 (2021)
17. Society, A.C.: Lung cancer statistics: how common is lung cancer (2020). https://www.cancer.org/cancer/lung-cancer/about/key-statistics.html

18. Svoboda, E.: Artificial intelligence is improving the detection of lung cancer. Nature **587**(7834), S20–S22 (2020)
19. Tam, L.K., Wang, X., Turkbey, E., Lu, K., Wen, Y., Xu, D.: Weakly supervised one-stage vision and language disease detection using large scale pneumonia and pneumothorax studies. In: Martel, A.L., Abolmaesumi, P., Stoyanov, D., Mateus, D., Zuluaga, M.A., Zhou, S.K., Racoceanu, D., Joskowicz, L. (eds.) MICCAI 2020. LNCS, vol. 12264, pp. 45–55. Springer, Cham (2020). https://doi.org/10.1007/978-3-030-59719-1_5
20. Ulyanov, D., Vedaldi, A., Lempitsky, V.: Instance normalization: the missing ingredient for fast stylization. arXiv preprint arXiv:1607.08022 (2016)
21. Xia, C., et al.: A Multi-modality Network for Cardiomyopathy Death Risk Prediction with CMR Images and Clinical Information. In: Shen, D., Liu, T., Peters, T.M., Staib, L.H., Essert, C., Zhou, S., Yap, P.-T., Khan, A. (eds.) MICCAI 2019. LNCS, vol. 11765, pp. 577–585. Springer, Cham (2019). https://doi.org/10.1007/978-3-030-32245-8_64
22. Yang, Z., Gong, B., Wang, L., Huang, W., Yu, D., Luo, J.: A fast and accurate one-stage approach to visual grounding. In: Proceedings of the IEEE/CVF International Conference on Computer Vision, pp. 4683–4693 (2019)
23. Yu, M., et al.: Toward rapid stroke diagnosis with multimodal deep learning. In: Martel, A.L., Abolmaesumi, P., Stoyanov, D., Mateus, D., Zuluaga, M.A., Zhou, S.K., Racoceanu, D., Joskowicz, L. (eds.) MICCAI 2020. LNCS, vol. 12263, pp. 616–626. Springer, Cham (2020). https://doi.org/10.1007/978-3-030-59716-0_59
24. Zhang, H., Gu, Y., Qin, Y., Yao, F., Yang, G.-Z.: Learning with sure data for nodule-level lung cancer prediction. In: Martel, A.L., Abolmaesumi, P., Stoyanov, D., Mateus, D., Zuluaga, M.A., Zhou, S.K., Racoceanu, D., Joskowicz, L. (eds.) MICCAI 2020. LNCS, vol. 12266, pp. 570–578. Springer, Cham (2020). https://doi.org/10.1007/978-3-030-59725-2_55

Fully Automatic Head and Neck Cancer Prognosis Prediction in PET/CT

Pierre Fontaine[1,2](✉), Vincent Andrearczyk[1], Valentin Oreiller[1,3],
Joël Castelli[2], Mario Jreige[3], John O. Prior[3], and Adrien Depeursinge[1,3]

[1] University of Applied Sciences and Arts Western Switzerland (HES-SO),
Sierre, Switzerland
pierre.fontaine@hevs.ch
[2] CLCC Eugene Marquis, INSERM, LTSI - UMR 1099, University Rennes,
35000 Rennes, France
[3] Centre Hospitalier Universitaire Vaudois (CHUV), Lausanne, Switzerland

Abstract. Several recent PET/CT radiomics studies have shown
promising results for the prediction of patient outcomes in Head and
Neck (H&N) cancer. These studies, however, are most often conducted
on relatively small cohorts (up to 300 patients) and using manually delineated tumors. Recently, deep learning reached high performance in the
automatic segmentation of H&N primary tumors in PET/CT. The automatic segmentation could be used to validate these studies on larger-scale
cohorts while obviating the burden of manual delineation. We propose
a complete PET/CT processing pipeline gathering the automatic segmentation of primary tumors and prognosis prediction of patients with
H&N cancer treated with radiotherapy and chemotherapy. Automatic
contours of the primary Gross Tumor Volume (GTVt) are obtained from
a 3D UNet. A radiomics pipeline that automatically predicts the patient
outcome (Disease Free Survival, DFS) is compared when using either
the automatically or the manually annotated contours. In addition, we
extract deep features from the bottleneck layers of the 3D UNet to compare them with standard radiomics features (first- and second-order as
well as shape features) and to test the performance gain when added to
them. The models are evaluated on the HECKTOR 2020 dataset consisting of 239 H&N patients with PET, CT, GTVt contours and DFS
data available (five centers). Regarding the results, using Hand-Crafted
(HC) radiomics features extracted from manual GTVt achieved the best
performance and is associated with an average Concordance (C) index of
0.672. The fully automatic pipeline (including deep and HC features from
automatic GTVt) achieved an average C index of 0.626, which is lower
but relatively close to using manual GTVt (p-value = 0.20). This suggests that large-scale studies could be conducted using a fully automatic
pipeline to further validate the current state of the art H&N radiomics.
The code will be shared publicly for reproducibility.

Keywords: Head and neck cancer · Radiomics · Deep learning

© Springer Nature Switzerland AG 2021
T. Syeda-Mahmood et al. (Eds.): ML-CDS 2021, LNCS 13050, pp. 59–68, 2021.
https://doi.org/10.1007/978-3-030-89847-2_6

1   Introduction

Radiomics allows quantitative analyses from radiological images with high throughput extraction to obtain prognostic patient information [1]. Radiomics features, including intensity-, texture-, and shape-based features, are generally extracted from Volumes Of Interests (VOI) that delineate the tumor volume [2]. These VOIs are often obtained from manual delineations made for treatment planning in radiotherapy [3,4] or from a simple threshold on the Positron Emission Tomography (PET) image [5]. This approach allowed researchers to perform various radiomics studies without the need for re-annotating the images specifically for these tasks. The annotations made for radiotherapy are, however, very large as compared to the true tumoral volumes and frequently include non-tumoral tissues and parts of other organs such as the trachea. On the other hand, simple PET thresholding may discard important hypo-metabolic parts of the tumoral volume (*e.g.* necrosis) containing important information for outcome prediction. For that reason, it seems important to clean delineations before using them in a radiomics study. Unfortunately, this represents a tedious and error-prone task when done manually, and usually entails poor inter-observer agreement [6]. Moreover, it is not scalable to large cohorts including thousands of patients, which is urgently needed to validate the true value of radiomics-based prognostic biomarkers when used in clinical routine.

Since the introduction of deep Convolutional Neural Network (CNN) segmentation models and in particular of the UNet architecture [7], the automatic segmentation of the tumoral volume in radiological and nuclear medicine images has made tremendous progress during the past ten years. In particular for brain tumor segmentation, expert-level performance was achieved in the context of the Brain Tumor Segmentation (BraTS) challenge [8,9]. The brain tumor segmentation models were further extended or used for prognostic prediction, allowing to obtain fully automatic methods that do not require manual delineation of VOIs for extracting the radiomics features. In particular, Baid *et al.* [10] used a 3D UNet for the segmentation task and then extracted radiomics features from the automatically segmented tumor components in multi-sequence Magnetic Resonance Imaging (MRI), achieving good performance for the prediction of overall patient survival. More recently, the automatic segmentation of Head and Neck (H&N) tumors from PET/CT imaging was investigated in the context of the HECKTOR challenge [6], also achieving expert-level delineation performance of the primary Gross Tumor Volume (GTVt). This opened avenues for scalable and fully automatic prognosis prediction in Head and Neck (H&N) cancer as well as their validation on large-scale cohorts.

In this paper, we propose a fully automatic pipeline for predicting patient prognosis (*i.e.* Disease-Free Survival, DFS) in oropharyngeal H&N cancer. We use a 3D UNet for segmenting the tumor contours and to extract deep radiomics features from its encoder. We further extract Hand-Crafted (HC) radiomics features from the automatically generated GTVt contours. Both PET and CT modalities are used to characterize the metabolic and morphological tumor tissue properties. Finally, we compare and combine the deep and HC radiomics features using either Cox Proportional Hazards (CPH) or Random Survival Forest (RSF) models. This fully automatic approach is compared to a classical radiomics approach based on HC features extracted from manually annotated VOIs.

2   Methods

This section first introduces the dataset. Then, the proposed manual and fully automatic radiomics pipelines are detailed. In particular, we compare the performance between prognostic radiomics models using either *manual* or *automatic* annotations of GTVt. We also compare the prognostic performance when using spatially aggregated activation maps from the bottleneck of the 3D UNet (*i.e.* the bottom of the encoder, referred to as *deep features*) and even combine these with HC radiomics features to investigate their complementarity for DFS prediction.

2.1   Datasets

For this study, we use the entire dataset used in the context of the HECKTOR 2020 challenge [6] (training and test sets) including a total of 239 patient with H&N cancer located in the oropharyngeal region (see Table 1). This dataset is an extension of the data used in the study of Vallières *et al.* 2017 [3].

Table 1. Overview of the dataset. Hôpital Général Juif (HGJ), Montréal, CA; Centre Hospitalier Universitaire de Sherbooke (CHUS), Sherbrooke, CA; Hôpital Maisonneuve-Rosemont (HMR), Montréal, CA; Centre Hospitalier de l'Université de Montréal (CHUM), Montréal; Centre Hospitalier Universitaire Vaudois (CHUV), CH.

Center	#Patients	Gender	Age (avg.)	Follow-up (avg. days)	#Events (DFS)
HGJ	55	Male 43 Female 12	62	1418	11
CHUS	71	Male 50 Female 21	62	1274	13
HMR	18	Male 14 Female 4	69	1265	4
CHUM	55	Male 41 Female 14	64	1120	7
CHUV	40	Male 35 Female 5	63	705	7

FluoroDeoxyGlucose-PET and Computed Tomography (FDG-PET/CT) imaging and manual GTVt contours are available for each patients. The latter (*i.e. manual*) were cleaned from initial VOIs acquired in the context of pre-treatment radiotherapy planning, *i.e.* re-annotated to specifically delineate the true tumoral volume. These manual GTVt are used both for the extraction of features in the *manual* setting and to train a 3D UNet that will generate *automatic* contours. The latter are generated for the entire dataset using a repeated 5-fold Cross-Validation (CV) scheme.

2.2 UNet Architecture and Training

In this section, we describe the deep learning model used to obtain *automatic* GTVt contours and *deep features*. We use a multi-modal 3D UNet developed in [11]. The input images are first cropped to $144 \times 144 \times 144$ voxels using the bounding boxes provided in the HECKTOR challenge [12]. The images are then resampled to $1\,mm^3$ with linear interpolation. The PET images are standardized to zero mean, unit variance. The CT images clipped to $[-1024,1024]$ and linearly mapped to $[-1,1]$. Data augmentation is applied to the training data including random shifts (maximum 30 voxels), sagittal mirroring and rotations (maximum $5°$). The model is trained with an Adam optimizer with cosine decay learning rate (initial learning rate 10^{-3}) to minimize a soft Dice loss. The model is trained for 600 epochs on the entire HECKTOR 2020 data (see Sect. 2.1). For each of the five folds of the repeated CV, the training set is split into 80% training and 20% validation. The best model according to the validation loss is used for obtaining both *automatic* contours and *deep features* on the test data. Since we do not have enough data to train the model on cases that are not present for the subsequent prognostic prediction study, we use the repeated 5-fold CV to mimic this scenario, ensuring that the *automatic* contours and *deep features* are all obtained on held-out data.

2.3 Radiomics Workflow

In order to evaluate and compare the DFS prognosis performance when extracting HC radiomics features from either the *manual* or *automatic* GTVt, as well as to evaluate the performance of the *deep features*, we develop a standard radiomics workflow, as detailed in Fig. 1.

We first pre-process the images (CT and PET) with an iso-resampling of $2 \times 2 \times 2$ mm using linear interpolation. This step is performed before feature extraction except for the shape features, for which we extract only from the CT with its initial resolution.

Then, we extract features from CT and PET images within either *manual* or *automatic* GTVt. Table 2 details the parameters and the types of HC features used. A total of 130 HC features are extracted per modality with additional 14 shape-based features. For each patient and for each annotation type, we therefore have 274 features as detailed in the following[1]. From those two modalities per patient (CT and PET), we extract features from the first-order family (18 features) and second-order family (56 features). Regarding the second-order family,

---

[1] We can unconventionally detail the number of HC features as follows: $274 = 2$ modalities $\times$ (2 bin $\times$ 56 2^{nd}order $+$ 18 1^{st}order) $+ 14$ shape (see Table 2).

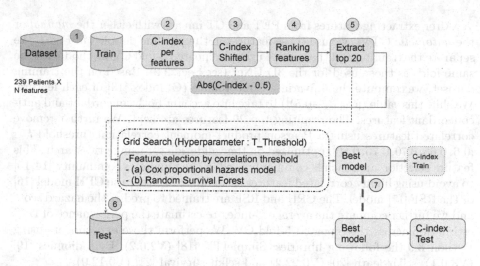

Fig. 1. Radiomics workflow inside one fold of the 5-fold CV and 1 repetition. The univariate and multivariate steps are highlighted in green and grey, respectively. (Color figure online)

we extract the 56 features using two different discretization parameters based on Fixed Bin Number (FBN) and Fixed Bin Size (FBS) (see Table 2). Those 56 features are divided into three sub-families: Grey Level Co-occurence Matrix (GLCM), Grey Level Run Length Matrix (GLRLM) and Grey Level Size Zone Matrix (GLSZM). Finally, we extract 14 shape features.

Table 2. List of the different combinations of parameters and types for HC radiomics features. FBN: Fixed Bin Number and FBS: Fixed Bin Size.

Image	Preprocessing	Settings	Features
CT	Iso-resampling 2 × 2 × 2 mm linear interpolation	FBN = 64 FBS = 50	GLCM (24 features) GLRLM (16 features) GLSZM (16 features)
			First Order (18 features)
			Shape (14 features)
PET	Iso-resampling 2 × 2 × 2 mm linear interpolation	FBN = 8 FBS = 1	GLCM (24 features) GLRLM (16 features) GLSZM (16 features)
			First Order (18 features)

Regarding the *deep features*, we extract a total of 256 ReLU-activated feature maps of size 9 × 9 × 9 at the UNet bottleneck, which are spatially averaged to obtain a collection of 256 scalar-valued features. The latter were directly used in the CPH or RSF model, either alone or concatenated with the HC features.

After extracting features from PET and CT images with either the *manual* or the *automatic* GTVt, we use those features in the pipeline detailed in Fig. 1. We separate the training and the test set using a repeated 5-fold CV method with the same folds as those used for the 3D UNet (see Sect. 2.2). Based on the training dataset, we compute the univariate Concordance (C) index [13] of each feature. We shift this value ($|C_{\text{index}} - 0.5|$) to take into account both concordant and anti-concordant features. The resulting top 20 features are kept. We further remove correlated features using a Pearson product-moment correlation threshold $t \in [0.6, 0.65, 0.70, 0.75, 0.80]$, optimized on the validation set using grid-search. This feature selection approach is commonly used in the radiomics community [14,15] to avoid using highly correlated features as an input to either the CPH model [16] or the RSF [17] model. The CPH and RSF are trained to predict the hazard score and we further compute the average C-index to estimate the performance of DFS estimation across the repeated 5-fold CV. We perform those experiments using Python 3.9 the following libraries: SimpleITK [18] (V2.0.2), PyRadiomics [19] (V3.0.1), scikit-learn [20] (V0.22.2) and scikit-survival [21] (V0.12.0).

3   Results

3.1   GTVt Segmentation

The estimated average Dice Score Coefficient (DSC) on the test sets of the repeated CV is 0.737 ± 0.009. It is worth noting that the model yielded a tumor contour for all patients in the dataset, *i.e.* at least one automatic GTV segmentation for each input. Figure 2 shows a qualitative result of the 3D UNet segmentation (*automatic* contour) compared with the *manual* contours.

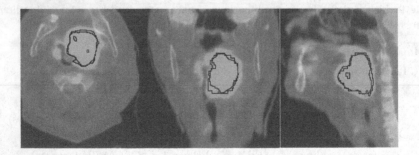

Fig. 2. Example of a *manual* GTVt in blue and the UNet-generated *automatic* GTVt in red on a fused PET/CT image. (Color figure online)

3.2   Prognosis Prediction

The DFS prediction performances are reported in Table 3 for the various approaches considered. For a baseline comparison, the performance of simple tumor volume-based DFS prediction is associated with a C-index of 0.6 for both manual and automatic GTVt.

Table 3. Comparison of prognosis performance (DFS) for the various approaches considered using either the CPH or the RSF model. We report the average C-index ± standard-error across 5 folds and 10 repetitions of the CV. The best overall performance is highlighted in bold and the best performance based on automatic segmentation is shown in italic.

	CPH	RSF
HC from *manual* GTVt	**0.672** ± 0.085	0.645 ± 0.096
HC from *automatic* GTVt	0.623 ± 0.111	0.623 ± 0.095
Deep features	0.560 ± 0.114	0.566 ± 0.108
Deep ft. + HC from *manual* GTVt	0.648 ± 0.094	0.639 ± 0.096
Deep ft. + HC from *automatic* GTVt	0.606 ± 0.027	*0.626* ± 0.028

4 Discussion and Conclusions

This work proposes a fully automatic processing pipeline for predicting patient prognosis from PET/CT in the context of H&N cancer. It is based on a 3D UNet providing automatic delineations of GTVt and deep features, as well as additional HC features extracted from the generated GTVt. Deep and HC features are compared and combined when used either by a CPH or RSF model predicting patient hazard for DFS. This fully automatic pipeline is compared to a classical approach based on manually delineated GTVt regions.

While being a secondary objective of our approach, the segmentation performance of the 3D UNet was found to be consistent with the top-performing models of the HECKTOR 2020 challenge [6]. While being close to the state of the art in the field, we will investigate ways to improve the segmentation performance in future work, which is expected to have a important impact on the DFS prediction performance.

The performance of prognostic prediction is reported in Sect. 3.2. According to Table 3, the best performance for DFS prediction is obtained with HC features from *manual* GTVt contours with a CPH model, which is associated with an average C-index across the repeated 5 folds CV of 0.672. Using the same HC features from *automatic* GTVt contours, results in a lower performance associated with C-indices of 0.623 and 0.623 using CPH or RSF, respectively. When used alone, the *deep features* obtain the worst prognosis performance with a C-index of either 0.560 (CPH) or 0.566 (RSF). Combining the deep features with HC extracted from *manual* GTVt seems to deteriorate the performance when compared to using HC features only in the *manual* VOIs (C-index of 0.648 versus 0.672 when based on CPH). By comparison, Andrearczyk *et al.* [22] developed a fully end-to-end model (segmentation and prediction) which achieved a good performance of 0.723 (C-index). Nevertheless, we can not directly compare this result with ours due to the difference in the validation splits. However, combining the *deep features* with HC extracted from *automatic* GTVt in a RSF model achieves a promising C-index of 0.626, where the rate of change is only 7.3% lower than the best model based on *manual* contours (CPH and HC features only) while being fully automatic. The statistical significance of this result (HC from *manual* GTVt versus deep ft. + HC from *automatic* GTVt) was assessed using a corrected r-repeated k-fold CV t-test as defined in Bouckaert *et al.* [23] and is associated with a p-value of 0.20. This correction is needed to adapt to underestimated variance related to dependent observations in repeated CVs. It is worth noting that using the standard uncorrected student's t-test yields a p-value of 0.0026. We feel that reporting the latter is important as many studies in the field are not using corrections when breaking the independance assumption of the t-test. Therefore, the proposed fully automatic prognostic model based on the combination of *deep features* with HC extracted from *automatic* GTVt with RSF model seems most relevant to conduct large-scale radiomics studies, which are crucial to further validate the clinical value of such image-based biomarkers. For inferring the prognosis of one patient, the computational time of the fully automatic pipeline (*i.e.* segmentation and *deep feature* extraction, HC features computation and prognosis prediction), is approximately 3 min (assuming that the 3D UNet is already trained). Moreover, the proposed approach allows scalable and reproducible contouring, as opposed to manual contouring where inter-observer agreement was reported to be low [6,24]. We remind the reader that we used a dataset originating from five distinct centers and including various scanner manufacturers and image acquisition protocols.

One limitation of our study is that the *deep features* are pooled by averaging the entire feature maps at the bottleneck of the UNet, which may explain the poor performance given by those set of features either alone or concatenated with the HC features. Future work will investigate feature pooling by averaging inside the VOIs (*e.g. automatic or manual*). Other aggregation methods will be investigated also for the HC features, as it was shown to have a strong influence on the prognosis predictive performance [25]. In addition, we plan to include readily available clinical patient data (*e.g.* age, gender, smoking status, tumor site, human papilloma virus status) [26] to further improve the prognosis performance of the models.

Acknowledgments. This work was partially supported by the Swiss National Science Foundation (SNSF, grant 205320_179069), the Swiss Personalized Health Network (SPHN via the IMAGINE and QA4IQI projects), and the Hasler Foundation (via the EPICS project, grant 20004).

References

1. Gillies, R.J., Kinahan, P.E., Hricak, H.: Radiomics: images are more than pictures, they are data. Radiology **278**(2), 563–577 (2016)
2. Zwanenburg, A., et al.: The image biomarker standardization initiative: standardized quantitative radiomics for high-throughput image-based phenotyping. Radiology **295**(2), 328–338 (2020)
3. Vallieres, M., et al.: Radiomics strategies for risk assessment of tumour failure in head-and-neck cancer. Sci. Rep. **7**(1), 1–14 (2017)

4. Andrearczyk, V., et al.: Automatic segmentation of head and neck tumors and nodal metastases in PET-CT scans. In International Conference on Medical Imaging with Deep Learning (MIDL) (2020)
5. Apostolova, I., et al.: Asphericity of pretherapeutic tumour FDG uptake provides independent prognostic value in head-and-neck cancer. Eur. Radiol. 24(9), 2077–2087 (2014)
6. Andrearczyk, V., et al.: Overview of the HECKTOR challenge at MICCAI 2020: automatic head and neck tumor segmentation in PET/CT. In: Lecture Notes in Computer Science (LNCS) Challenges (2021)
7. Ronneberger, O., Fischer, P., Brox, T.: U-Net: Convolutional Networks for Biomedical Image Segmentation. In: Navab, N., Hornegger, J., Wells, W.M., Frangi, A.F. (eds.) MICCAI 2015. LNCS, vol. 9351, pp. 234–241. Springer, Cham (2015). https://doi.org/10.1007/978-3-319-24574-4_28
8. Menze, B.H., et al.: The multimodal brain tumor image segmentation benchmark (BRATS). IEEE Trans. Med. Imaging 34(10), 1993–2024 (2014)
9. Havaei, M.: Brain tumor segmentation with deep neural networks. Med. Image Anal. 35, 18–31 (2017)
10. Baid, U., et al.: Deep Learning Radiomics Algorithm for Gliomas (DRAG) Model: A Novel Approach Using 3D UNET Based Deep Convolutional Neural Network for Predicting Survival in Gliomas. In: Crimi, A., Bakas, S., Kuijf, H., Keyvan, F., Reyes, M., van Walsum, T. (eds.) BrainLes 2018. LNCS, vol. 11384, pp. 369–379. Springer, Cham (2019). https://doi.org/10.1007/978-3-030-11726-9_33
11. Isensee, F., Kickingereder, P., Wick, W., Bendszus, M., Maier-Hein, K.H.: Brain tumor segmentation and radiomics survival prediction: contribution to the BRATS 2017 challenge. In: Crimi, A., Bakas, S., Kuijf, H., Menze, B., Reyes, M. (eds.) BrainLes 2017. LNCS, vol. 10670, pp. 287–297. Springer, Cham (2018). https://doi.org/10.1007/978-3-319-75238-9_25
12. Andrearczyk, V., Oreiller, V., Depeursinge, A.: Oropharynx detection in PET-CT for tumor segmentation. In: Irish Machine Vision and Image Processing (2020)
13. Harrell Jr, F.E., Lee, K.L., Mark, D.B.: Multivariable prognostic models: issues in developing models, evaluating assumptions and adequacy, and measuring and reducing errors. Stat. Med. 15(4), 361–387 (1996)
14. Lambin, P., et al.: Radiomics: the bridge between medical imaging and personalized medicine. Nat. Rev. Clin. Oncol. 14(12), 749–762 (2017)
15. Suter, Y., et al.: Radiomics for glioblastoma survival analysis in pre-operative MRI: exploring feature robustness, class boundaries, and machine learning techniques. Cancer Imaging 20(1), 1–13 (2020)
16. David, C.R., et al.: Regression models and life tables (with discussion). J. R. Stat. Soc. 34(2), 187–220 (1972)
17. Ishwaran, H., et al.: Random survival forests. Ann. Appl. Stat. 2(3), 841–860 (2008)
18. Lowekamp, B.C., Chen, D.T., Ibáñez, L., Blezek, D.: The design of simpleitk. Front. Neuroinf. 7, 45 (2013)
19. Van Griethuysen, J.J., et al.: Computational radiomics system to decode the radiographic phenotype. Cancer Res. 77(21), e104–e107 (2017)
20. Pedregosa, F., et al.: Scikit-learn: machine learning in Python. J. Mach. Learn. Res. 12, 2825–2830 (2011)
21. Pölsterl, S.: scikit-survival: a library for time-to-event analysis built on top of scikit-learn. J. Mach. Learn. Res. 21(212), 1–6 (2020)

22. Andrearczyk, V., et al.: Multi-task deep segmentation and radiomics for automatic prognosis in head and neck cancer. In: Rekik, I., Adeli, E., Park, S.H., Schnabel, J. (eds.) Predictive Intelligence in Medicine. PRIME 2021. Lecture Notes in Computer Science, vol. 12928, Springer, Cham (2021). https://doi.org/10.1007/978-3-030-87602-9_14
23. Bouckaert, R.R., Frank, E.: Evaluating the replicability of significance tests for comparing learning algorithms. In: Dai, H., Srikant, R., Zhang, C. (eds.) PAKDD 2004. LNCS (LNAI), vol. 3056, pp. 3–12. Springer, Heidelberg (2004). https://doi.org/10.1007/978-3-540-24775-3_3
24. Vorwerk, H., et al.: The delineation of target volumes for radiotherapy of lung cancer patients. Radiother. Oncol. **91**(3), 455–460 (2009)
25. Fontaine, P., Acosta, O., Castelli, J., De Crevoisier, R., Müller, H., Depeursinge, A.: The importance of feature aggregation in radiomics: a head and neck cancer study. Sci. Rep. **10**(1), 1–11 (2020)
26. Zhai, T.T., et al.: Improving the prediction of overall survival for head and neck cancer patients using image biomarkers in combination with clinical parameters. Radiother. Oncol. **124**(2), 256–262 (2017)

Feature Selection for Privileged Modalities in Disease Classification

Winston Zhang[1]([✉])(iD), Najla Al Turkestani[1], Jonas Bianchi[1,2,3], Celia Le[1],
Romain Deleat-Besson[1], Antonio Ruellas[1], Lucia Cevidanes[1], Marilia Yatabe[1],
Joao Gonçalves[2], Erika Benavides[1], Fabiana Soki[1], Juan Prieto[4],
Beatriz Paniagua[4], Jonathan Gryak[1], Kayvan Najarian[1],
and Reza Soroushmehr[1]

[1] University of Michigan, Ann Arbor, MI, USA
wwzhang@umich.edu
[2] São Paulo State University, São Paulo, Brazil
[3] University of the Pacific, San Francisco, CA, USA
[4] University of North Carolina, Chapel Hill, NC, USA

Abstract. Multimodal data allows supervised learning while considering multiple complementary views of a problem, improving final diagnostic performance of trained models. Data modalities that are missing or difficult to obtain in clinical situations can still be incorporated into model training using the learning using privileged information (LUPI) framework. However, noisy or redundant features in the privileged modality space can limit the amount of knowledge transferred to the diagnostic model during the LUPI learning process. We consider the problem of selecting desirable features from both standard features which are available during both model training and testing, and privileged features which are only available during model training. A novel filter feature selection method named NMIFS+ is introduced that considers redundancy between standard and privileged feature spaces. The algorithm is evaluated on two disease classification datasets with privileged modalities. Results demonstrate an improvement in diagnostic performance over comparable filter selection algorithms.

Keywords: Privileged learning · Mutual information · Knowledge transfer · Feature selection · Multimodal data · Clinical decision support

1 Introduction

Data about an event or system acquired from multiple different types of conditions, instruments, or techniques is termed multimodal data, and each separate data domain is termed a "modality" [12]. By integrating complementary views or modalities of an observed source through data fusion, a more complete representation of the observed source can be developed [3].

Grant supported by NIDCR R01 DE024450.

In applications such as action recognition, clinical imaging diagnosis, and self-driving cars, techniques for integrating multimodal data have included transfer learning [16], data distillation [9], and semi-supervised learning [6]. Data fusion has proven especially useful in training models for imaging applications [6]. However, in clinical settings some data modalities may be difficult to collect or unavailable during the model testing stage [13]. The availability of data domains such as blood protein levels or additional radiographic scans may be limited by time or cost during patient diagnosis. A way to transfer multimodal information to a model that can still perform single-modal clinical analysis is needed.

The learning using privileged information (LUPI) paradigm [23] uses an additional data domain termed "privileged data" during the model training process that is not available during testing. Privileged data can provide knowledge such as the relative "hardness" of individual training samples to the trained model [22]. The LUPI process can be equated to a student-teacher example, where the privileged domain acts as a teacher that guides the student model to find a more generalizable representation of the underlying population. During model testing, the teacher is no longer available to help the student.

LUPI has been previously applied to multimodal learning in clinical applications. Li et al. has used LUPI-based models to integrate MRI and PET neuroimaging data for Alzheimer's disease classification [13]. Ye et al. graded glioma images using privileged learning [24]. Duan et al. used genetic information as a privileged domain for single-modal imaging detection of glaucoma [7].

Although the LUPI paradigm provides a method of integrating multimodal data even under conditions of missing modalities during testing, it also introduces a new consideration: the quality of the information transferred by the privileged teacher. Traditional feature selection aims to choose an ideal subset of features that are both relevant to the training label, and non-redundant so as to reduce noise and dimensionality of the trained model [5,10,19]. In LUPI, two subsets of features must be chosen from the standard and privileged feature sets respectively during the training process, of which only the standard feature subset will be used during the testing phase. An ideal privileged feature subset is not only relevant to the problem at hand but also non-redundant relative to the standard feature subset.

We propose a new filter feature selection method that takes into account the relationship between the standard and privileged sets during the LUPI training process. The algorithm aims to improve the knowledge transferred from the privileged modality during the LUPI paradigm. We then test the proposed algorithm on two different multimodal disease classification datasets with several commonly used LUPI-based classifiers.

2   Background

2.1   Learning Using Privileged Information

In non-LUPI binary classification problems, we are given a set of training pairs generated from an underlying distribution:

$$\Gamma = \{(x_i^t, y_i) | x_i^t \in X^t, \ y_i \in Y, \ ||x_i^t||_0 = d_t, \ i = 1...n\} \tag{1}$$

where n is the number of training samples, X^t contains the standard modality of the training samples, F^t represents the standard feature set with d_t number of features, and Y represents the set of training labels.

In the LUPI framework, the training set can be represented as a set of triplets:

$$\Gamma = \{(x_i^t, x_i^p, y_i) | x_i^t \in X^t, \ x_i^p \in X^p, \ y_i \in Y, \ ||x_i^t||_0 = d_t, \ ||x_i^p||_0 = d_p, \ i = 1...n\} \tag{2}$$

where X^p contains the privileged modality of the training samples, F^p represents the privileged feature set, and d_p is the number of features in F^p [23].

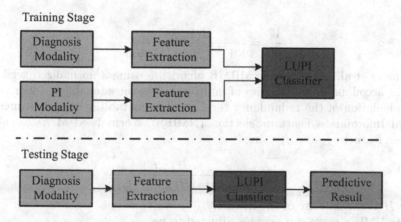

Fig. 1. Diagram of training and testing process in the LUPI framework [13].

Figure 1 describes the steps for training and testing a LUPI-based clinical model. The standard feature set, F^t, and the privileged feature set, F^p, undergo feature selection to find two new subsets of features S^t and S^p for classifier training. A LUPI model is then trained on the training samples with standard feature subset S^t and labels Y, while using S^p as the privileged feature subset for knowledge transfer.

In the testing stage, the privileged modality X^p is unavailable. The feature subset S^t previously selected from F^t during training is used for single-modal classifier inference on the unseen test set samples. A classification is obtained for each test sample.

2.2 Mutual Information Feature Selection

Feature selection methods are generally divided into filter and wrapper methods [5]. Many popular filter algorithms use the mutual information (MI) criteria as a measure of both relevancy and redundancy of features, due to its capability of quantifying nonlinear relationships between variables and its invariance to different transformations [11]. If X and Y are two continuous random variables with joint distribution $p(x, y)$ and marginal distributions $p(x)$ and $p(y)$ respectively, we can calculate the MI between them as:

$$I(X; Y) = \int \int p(x, y) log \frac{p(x, y)}{p(x)p(y)} dx dy \tag{3}$$

Given feature set F with d features, n samples, and labels Y, Maximum Relevancy Minimum Redundancy (MRMR) is a widely used incremental search filter algorithm that uses MI to select the optimal feature set S with cardinality $|S| = k$ as a hyperparameter [8,19]. MRMR calculates the relevancy V_i by finding the MI between a feature $f_i \in F$ and label Y. The redundancy W_i is calculated between feature f_i and every feature already chosen in set S. Features are then ranked and chosen by maximizing MRMR score produced as a ratio between V_i and W_i:

$$\frac{V_i}{W_i} = \frac{I(Y; f_i)}{|S|^{-1} \sum_{f_s \in S} I(f_i; f_s)} \tag{4}$$

Estévez et al. improved the MRMR algorithm using a normalized version of MI that accounts for the MI bias of multivalued features, enabling a more accurate calculation of the redundancy term [8]. The method is named Normalized Mutual Information Feature Selection (NMIFS). Normalized MI is calculated between features f_i and f_s using

$$NI(f_i; f_s) = \frac{I(f_i; f_s)}{min\{H(f_i), H(f_s)\}} \tag{5}$$

where H is the entropy of the feature.

The NMIFS score can now be adjusted to be

$$\frac{V_i}{W_i} = \frac{I(Y; f_i)}{|S|^{-1} \sum_{f_s \in S} NI(f_i; f_s)} \tag{6}$$

3 Method

We propose an extension of the NMIFS algorithm to the LUPI case named NMIFS+. Consider the standard features F^t and privileged features F^p. Applying a feature selection algorithm would result in feature sets S^t and S^p containing selected features from F^t and F^p respectively.

A simple method of selecting sets S^t and S^p would be to apply NMIFS separately on F^t and F^p, as shown in Fig. 2. The NMIFS score for each feature

$f_{i,t} \in F^t$ considers $V_{i,t}$ (the relevancy of $f_{i,t}$ compared to label Y) and $W_{i,t}$ (the redundancy of $f_{i,t}$ compared to every feature $f_{s,t}$ already contained in S^t). The corresponding relevancy $V_{i,p}$ and redundancy $W_{i,p}$ terms are considered for each $f_{i,p} \in F^p$.

In the LUPI paradigm, however, the relationship between the chosen standard set S^t and the privileged set S^p must be considered. We wish to select for features in S^p that are non-redundant with the standard set S^t, since the goal of LUPI is to transfer useful information between S^p and S^t during the training stage. If S^p contains noisy, irrelevant, or redundant information, knowledge transferred may be detrimental to the quality of the final model.

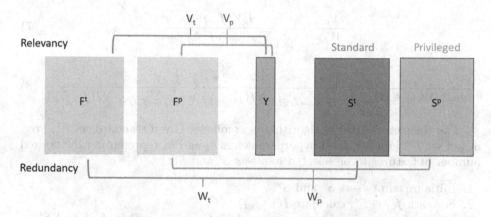

Fig. 2. Visualization of relevancy and redundancy terms considered in NMIFS applied separately to F^t and F^p.

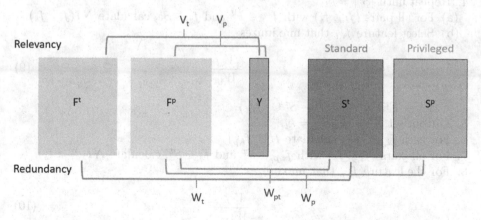

Fig. 3. Visualization of relevancy and redundancy terms considered in NMIFS+.

In the proposed method NMIFS+, as seen in Fig. 3, an extra redundancy term W_{pt} is added that considers the redundancy between standard feature set S^t and yet to be chosen privileged features in F^p. NMIFS+ selects the standard feature set S^t first before selecting the feature set S^p in order to ensure that the redundancy between S^t and S^p is minimized. We hypothesize that by penalizing redundancy between the selected standard set and yet to be selected privileged features, the transfer of redundant information from S^p to S^t during LUPI training will decrease, resulting in a more useful privileged modality teaching process.

From Eq. 6, we obtain the two separate NMIFS+ criteria for selecting features from F^t and F^p respectively:

$$\frac{V_{i,t}}{W_{i,t}} = \frac{I(Y; f_{i,t})}{\frac{1}{|S^t|}\sum_{f_s \in S^t} NI(f_s; f_{i,t})} \tag{7}$$

$$\frac{V_{i,p}}{\text{mean}(W_{i,pt} + W_{i,p})} = \frac{2 * I(Y; f_{i,p})}{\frac{1}{|S^t|}\sum_{f_s \in S^t} NI(f_s; f_{i,p}) + \frac{1}{|S^p|}\sum_{f_s \in S^p} NI(f_s; f_{i,p})} \tag{8}$$

The complete NMIFS+ algorithm is as follows. Given standard set F^t, privileged set F^p, labels Y, and hyperparameters k_t and k_p representing the desired number of features to be selected into sets S^t and S^p:

1. Initialize empty sets S^t and S^p.
2. For each $f_{i,t} \in F^t$, calculate $I(Y; f_{i,t})$.
3. For the feature $f_{i,t}$ that maximizes $I(Y; f_{i,t})$, $F^t \leftarrow F^t \setminus \{f_{i,t}\}$ and $S^t \leftarrow S^t \cup \{f_{i,t}\}$
4. Repeat until $|S^t| = k_t$.
   (a) For all pairs $(f_{i,t}, f_s)$ with $f_{i,t} \in F^t$ and $f_s \in S^t$, calculate $NI(f_{i,t}; f_s)$.
   (b) Select feature $f_{i,t}$ that maximizes

$$\frac{V_{i,t}}{W_{i,t}} \tag{9}$$

   $F^t \leftarrow F^t \setminus \{f_{i,t}\}$. $S^t \leftarrow S^t \cup \{f_{i,t}\}$.
5. Output set S^t with $|S^t| = k_t$.
6. For each $f_{i,p} \in F^p$, calculate $I(Y; f_{i,p})$.
7. For all pairs $(f_{i,p}, f_s)$ with $f_{i,p} \in F^p$ and $f_s \in S^t$, calculate $NI(f_{i,p}; f_s)$.
8. For the feature $f_{i,p}$ that maximizes

$$\frac{V_{i,t}}{W_{i,p}} \tag{10}$$

   $F^p \leftarrow F^p \setminus \{f_{i,p}\}$ and $S^p \leftarrow S^p \cup \{f_{i,p}\}$
9. Repeat until $|S^p| = k_p$.
   (a) For all pairs $(f_{i,p}, f_s)$ with $f_{i,p} \in F^p$ and $f_s \in S^p$, calculate $NI(f_{i,p}; f_s)$.
   (b) Select feature $f_{i,p}$ that maximizes

$$\frac{V_{i,p}}{\text{mean}(W_{i,pt} + W_{i,p})} \tag{11}$$

$F^p \leftarrow F^p \setminus \{f_{i,p}\}. \ S^p \leftarrow S^p \cup \{f_{i,p}\}.$

10. Output set S^p with $|S^p| = k_p$.

During the testing stage, only feature set S^t is used for model evaluation. When implementing NMIFS+, we calculate MI for continuous features using histograms with bin number of $\sqrt{n}$, where n is the number of training samples. MI for discrete features is found using a contingency table.

4 Experiments

We evaluate three different feature selection algorithms for LUPI: 1) simple ranking of features by Area Under Receiver Operator Curve (AUC) with the target label, 2) NMIFS applied separately on standard and privileged sets, and 3) NMIFS+.

We use an 80%–20% train-test split for all datasets. A grid search and 5-fold cross-validation (CV) is performed on the training set to determine the optimal parameters for each compared LUPI model. Parameters k_t and k_p are also grid searched over the range $[5, 20]$ and $[5, 20]$ respectively. Test results are obtained by taking the average of test set predictions from all 5 models trained during the best 5-fold CV run (based on validation performance). The overall procedure is repeated 10 times with 10 random seeds to avoid sampling bias from random train-test partitioning. Final reported results are the averaged mean ± standard deviation (SD) test set performances across all 10 repetitions.

4.1 Compared LUPI Models

We test our algorithm with two LUPI-based models, SVM+ and KRVFL+. We also include comparisons with the corresponding non-LUPI models SVM and RVFL using regular AUC and NMIFS feature selection.

SVM and SVM+. The support vector machine with gaussian kernel has been used frequently for machine learning classification. SVM+, first proposed by Vapnik et al. [23], uses the privileged space to model slack variables found in soft-margin SVM. We grid search for the hyperparameters γ and C over the ranges [10e–3, 10e3]. The alternating SMO algorithm introduced by Pechyony et al. is used for optimization [18].

RVFL and KRVFL+. The random vector functional link network (RVFL) [17] is a shallow neural network with randomized network elements. It incorporates the LUPI framework by using the kernel-based RVFL+ algorithm (KRVFL+), which approximates SVM+ with fewer optimization constraints [25]. The hyperparameters C, τ, and γ are searched for over the ranges [1, 2], [0.01, 0.03], and [1, 5000] respectively following paper suggested ranges.

4.2   Datasets

Parkinson's Disease Dataset. A real-world dataset for classifying Parkinson's Disease from patients' vocal phonations [20] is taken from the UCI machine learning repository [14]. The dataset consists of 188 positive and 64 control patients, with voice samples from each patient averaged to form one representative sample. Baseline, time frequency, and vocal fold features are used as standard features while Mel Frequency Cepstral Coefficient (MFCC) features are used as the privileged modality, giving a total of 54 standard and 84 privileged features.

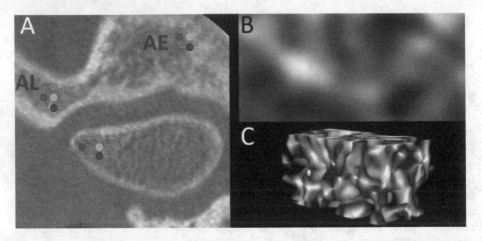

Fig. 4. Selection of the volume of interests (VOIs) (A) Axial view depicting VOIs in the condyle, the articular eminence (AE), and the anterolateral (AL) surface of the articular fossa. (B) Example of an extracted VOI containing the radiomics and bone morphometry data (C) 3-dimensional rendering of an extracted VOI.

Osteoarthritis Dataset. The dataset includes collected data from 46 early-stage temporomandibular joint (TMJ) osteoarthritis (OA) patients and 46 age and gender matched control patients. OA diagnosis was verified by a TMJ specialist using the Diagnostic Criteria for Temporomandibular Disorders (DC/TMD) [21]. The study followed guidelines of the Institutional Review Board (IRB) at the University of Michigan (number HUM00113199).

The standard modality consists of 54 radiomic and 6 clinical features. From the TMJ region, 18 radiomic features were extracted from the articular eminence, 14 from the anterolateral fossa surface, and 24 from the condyle region. Extracted volumes were histogram matched to surrounding areas for normalization of feature values. A visualization is found in Fig. 4. Imaging procedure details can be found in [1] and [2].

The privileged modality consists of 25 protein level features collected from blood and saliva. The features were previously found to be correlated with TMJ disease progression [4]. Acquisition procedure is described in [2]. Protein-protein interactions were also previously found to be significant to OA diagnosis [2], so $\sum_{i=1}^{25} i = 325$ new pairwise product protein features are generated, resulting in 350 privileged features in total. Protein data are expensive to obtain for patients, making the features a good choice for a privileged modality.

5 Results

5.1 Parkinson's Dataset

The class imbalance in this dataset means that AUPRC is a better measure of model performance [15]. AUPRC baseline is $\frac{188}{252} = 0.746$. From Table 1, we see that LUPI performs better than non-LUPI in terms of AUPRC across all models. NMIFS+ generally increases mean performance and decreases variance for both SVM+ and KRVFL+ when compared to NMIFS and AUC.

Table 1. Parkinson's performance comparison

Algorithm	AUC	AUPRC	F1	Accuracy
SVM [AUC]	0.593 ± 0.067	0.760 ± 0.050	$\mathbf{0.789 \pm 0.064}$	$\mathbf{70.9 \pm 5.5}$
SVM [NMIFS]	0.628 ± 0.054	0.783 ± 0.037	0.780 ± 0.077	70.3 ± 6.7
RVFL [AUC]	0.661 ± 0.067	$\mathbf{0.814 \pm 0.042}$	0.728 ± 0.094	66.1 ± 8.1
RVFL [NMIFS]	$\mathbf{0.664 \pm 0.069}$	0.807 ± 0.042	0.767 ± 0.089	69.5 ± 8.3
SVM+ [AUC]	0.723 ± 0.043	0.860 ± 0.028	0.712 ± 0.084	65.4 ± 6.7
SVM+ [NMIFS]	0.729 ± 0.044	0.855 ± 0.027	0.735 ± 0.095	67.6 ± 7.8
SVM+ [NMIFS+]	$\mathbf{0.736 \pm 0.033}$	$\mathbf{0.863 \pm 0.019}$	$\mathbf{0.758 \pm 0.072}$	$\mathbf{69.2 \pm 6.0}$
KRVFL+ [AUC]	0.698 ± 0.045	0.841 ± 0.031	0.728 ± 0.092	66.7 ± 7.5
KRVFL+ [NMIFS]	0.710 ± 0.043	0.846 ± 0.026	0.734 ± 0.081	66.9 ± 6.9
KRVFL+ [NMIFS+]	0.733 ± 0.039	0.854 ± 0.025	0.753 ± 0.084	69.0 ± 7.5

5.2 TMJ Osteoarthritis Dataset

The TMJ dataset is balanced, meaning that AUC and AUPRC are well correlated. In Table 2 we see a similar increase in performance from the introduction of the privileged modality as in the Parkinson's dataset. Although variance was similar between NMIFS and NMIFS+, mean performance is improved using the NMIFS+ method.

Table 2. TMJ performance comparison

Algorithm	AUC	F1	Accuracy
SVM [AUC]	0.771 ± 0.083	0.710 ± 0.067	73.1 ± 5.6
SVM [NMIFS]	**0.792 ± 0.069**	**0.727 ± 0.090**	**74.3 ± 5.8**
RVFL [AUC]	0.761 ± 0.060	0.715 ± 0.056	73.1 ± 5.3
RVFL [NMIFS]	0.778 ± 0.059	0.707 ± 0.076	73.2 ± 5.6
SVM+ [AUC]	0.794 ± 0.065	0.733 ± 0.062	74.1 ± 4.5
SVM+ [NMIFS]	0.821 ± 0.047	0.737 ± 0.062	75.6 ± 4.6
SVM+ [NMIFS+]	0.823 ± 0.045	0.742 ± 0.063	76.1 ± 4.6
KRVFL+ [AUC]	0.820 ± 0.042	0.756 ± 0.057	77.4 ± 4.2
KRVFL+ [NMIFS]	0.827 ± 0.029	0.771 ± 0.040	78.3 ± 3.4
KRVFL+ [NMIFS+]	**0.841 ± 0.036**	**0.780 ± 0.036**	**79.3 ± 2.9**

In Fig. 5, the number of standard features selected k_t was held at a constant value of 5, while the number of privileged features varied in the range [2, 20]. NMIFS and NMIFS+ choose the same 5 standard features since their only difference is during privileged selection. NMIFS+ consistently chooses a more informative privileged subset and converges to an optimal subset quickly. AUC chooses a different set of standard features, leading to low performance.

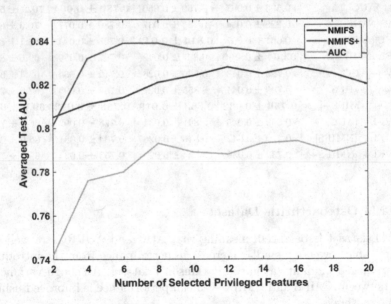

Fig. 5. Comparison of feature selection performance using KRVFL+ on the TMJ dataset. Y-axis represents the averaged test set AUC performance over 10 random splits. The number of standard features selected k_t is fixed to 5, while the number of privileged features selected k_p is varied.

6   Conclusions

We developed an extension of the classic feature selection algorithm MRMR to the LUPI paradigm and showed that accounting for redundancy between standard and privileged modalities can help improve final LUPI model performance. Different views of the same clinical event, such as protein levels and radiographic data from patients with TMJ OA, can be cleaned of redundant information and integrated to train a better-performing single-modal model.

More experiments can be performed to determine if NMIFS+ performs better on privileged modalities with high levels of redundancy with the standard feature set. Further optimization of the NMIFS+ algorithm is also possible by adjusting for the amount of contribution between the standard and privileged redundancy terms during privileged feature selection.

References

1. Bianchi, J., et al.: Software comparison to analyze bone radiomics from high resolution CBCT scans of mandibular condyles. Dentomaxill. Radiol. **48**(6), 20190049 (2019)
2. Bianchi, J., et al.: Osteoarthritis of the temporomandibular joint can be diagnosed earlier using biomarkers and machine learning. Sci. Rep. **10**(1), 1–14 (2020)
3. Bleiholder, J., Naumann, F.: Data fusion. ACM Comput. Surv. (CSUR) **41**(1), 1–41 (2009)
4. Cevidanes, L.H., et al.: 3d osteoarthritic changes in tmj condylar morphology correlates with specific systemic and local biomarkers of disease. Osteoarth. Cart. **22**(10), 1657–1667 (2014)
5. Chandrashekar, G., Sahin, F.: A survey on feature selection methods. Comput. Electr. Eng. **40**(1), 16–28 (2014)
6. Cheplygina, V., de Bruijne, M., Pluim, J.P.: Not-so-supervised: a survey of semi-supervised, multi-instance, and transfer learning in medical image analysis. Med. Image Anal. **54**, 280–296 (2019)
7. Duan, L., et al.: Incorporating privileged genetic information for fundus image based glaucoma detection. In: Golland, P., Hata, N., Barillot, C., Hornegger, J., Howe, R. (eds.) MICCAI 2014. LNCS, vol. 8674, pp. 204–211. Springer, Cham (2014). https://doi.org/10.1007/978-3-319-10470-6_26
8. Estévez, P.A., Tesmer, M., Perez, C.A., Zurada, J.M.: Normalized mutual information feature selection. IEEE Trans. Neural Netw. **20**(2), 189–201 (2009)
9. Hinton, G., Vinyals, O., Dean, J.: Distilling the knowledge in a neural network. arXiv preprint arXiv:1503.02531 (2015)
10. Izmailov, R., Lindqvist, B., Lin, P.: Feature selection in learning using privileged information. In: 2017 IEEE International Conference on Data Mining Workshops (ICDMW), pp. 957–963. IEEE (2017)
11. Kullback, S.: Information theory and statistics. Courier Corporation (1997)
12. Lahat, D., Adali, T., Jutten, C.: Multimodal data fusion: an overview of methods, challenges, and prospects. Proc. IEEE **103**(9), 1449–1477 (2015)
13. Li, Y., Meng, F., Shi, J.: Learning using privileged information improves neuroimaging-based cad of Alzheimer's disease: a comparative study. Med. Biol. Eng. Comput. **57**(7), 1605–1616 (2019)

14. Lichman, M., et al.: Uci machine learning repository (2013)
15. Ozenne, B., Subtil, F., Maucort-Boulch, D.: The precision-recall curve overcame the optimism of the receiver operating characteristic curve in rare diseases. J. Clin. Epidemiol. **68**(8), 855–859 (2015)
16. Pan, S.J., Yang, Q.: A survey on transfer learning. IEEE Trans. Knowl. Data Eng. **22**(10), 1345–1359 (2009)
17. Pao, Y.H., Park, G.H., Sobajic, D.J.: Learning and generalization characteristics of the random vector functional-link net. Neurocomputing **6**(2), 163–180 (1994)
18. Pechyony, D., Izmailov, R., Vashist, A., Vapnik, V.: SMO-style algorithms for learning using privileged information. In: Dmin. pp. 235–241. Citeseer (2010)
19. Peng, H., Long, F., Ding, C.: Feature selection based on mutual information criteria of max-dependency, max-relevance, and min-redundancy. IEEE Trans. Patt. Anal. Mach. Intell. **27**(8), 1226–1238 (2005)
20. Sakar, C.O., et al.: A comparative analysis of speech signal processing algorithms for Parkinson's disease classification and the use of the tunable q-factor wavelet transform. Appl. Soft Comput. **74**, 255–263 (2019)
21. Schiffman, E., et al.: Diagnostic criteria for temporomandibular disorders (DC/TMD) for clinical and research applications: recommendations of the international RDC/TMD consortium network and orofacial pain special interest group. J. Oral Facial Pain Head. **28**(1), 6 (2014)
22. Sharmanska, V., Quadrianto, N., Lampert, C.H.: Learning to rank using privileged information. In: Proceedings of the IEEE International Conference on Computer Vision, pp. 825–832 (2013)
23. Vapnik, V., Vashist, A.: A new learning paradigm: learning using privileged information. Neural Netw. **22**(5–6), 544–557 (2009)
24. Ye, F., Pu, J., Wang, J., Li, Y., Zha, H.: Glioma grading based on 3d multimodal convolutional neural network and privileged learning. In: 2017 IEEE International Conference on Bioinformatics and Biomedicine (BIBM), pp. 759–763. IEEE (2017)
25. Zhang, P.B., Yang, Z.X.: A new learning paradigm for random vector functional-link network: RVFL+. Neural Netw. **122**, 94–105 (2020)

Merging and Annotating Teeth and Roots from Automated Segmentation of Multimodal Images

Romain Deleat-Besson[1], Celia Le[1], Winston Zhang[1], Najla Al Turkestani[1],
Lucia Cevidanes[1(✉)], Jonas Bianchi[5], Antonio Ruellas[1], Marcela Gurgel[1],
Camila Massaro[1], Aron Aliaga Del Castillo[1], Marcos Ioshida[1],
Marilia Yatabe[1], Erika Benavides[1], Hector Rios[1], Fabiana Soki[1], Gisele Neiva[1],
Kayvan Najarian[1], Jonathan Gryak[1], Martin Styner[2],
Juan Fernando Aristizabal[3], Diego Rey[4], Maria Antonia Alvarez[4], Loris Bert[6],
Reza Soroushmehr[1], and Juan Prieto[2]

[1] University of Michigan, Ann Arbor, MI, USA
luciacev@umich.edu
[2] University of North Carolina, Chapel Hill, NC, USA
[3] University of Valle, Cali, Colombia
[4] University CES, Medellin, Colombia
[5] University of the Pacific, San Francisco, CA, USA
[6] New York University, New York, NY, USA

Abstract. This paper aims to combine two different imaging techniques to create an accurate 3D model representation of root canals and dental crowns. We combine Cone-Beam Computed Tomography (CBCT) (root canals) and Intra Oral Scans (IOS) (dental crowns). The Root Canal Segmentation algorithm relies on a U-Net architecture with 2D sliced images from CBCT scans as its input. The segmentation task achieved an F1-score of 0.84. The IOS segmentation (Dental Model Segmentation) algorithm and Universal Labeling and Merging (ULM) algorithm use a multi-view approach for 3D shape analysis. The approach consists of acquiring views of the 3D object from different viewpoints and extract surface features such as the normal vectors. The generated 2D images are then analyzed via a 2D convolutional neural networks (U-Net) for segmentation or classification tasks. The segmentation task on IOS achieved an accuracy of 0.9. The ULM algorithm classifies the jaws between upper and lower and aligns them to a template and labels each crown and root with the 'Universal Numbering System' proposed by the 'American Dental Association'. The ULM task achieve an F1-score of 0.85. Merging and annotated of CBCT and IOS imaging modalities will help guide clinical decision support and quantitative treatment planning for specific teeth, implant placement, root canal treatment, restorative procedures, or biomechanics of tooth movement in orthodontics.

Keywords: Deep learning · Root canal segmentation · Dental crown segmentation · Universal label · Merging · Dentistry

Supported by NIDCR R01 DE024450.

1   Introduction

Comprehensive dentistry treatments aim to keep the integrity of soft tissues, jaw bones, and teeth. It requires the acquisition of different diagnostic imaging modalities such as periapical x-rays, Cone Beam Computed Tomography (CBCT), panoramic radiography, cephalogram [33], or more recent modalities such as Intra Oral Scan (IOS) scanners which are used to generate digital dental models and make ceramic dental crowns [4]. All these imaging techniques have made the complex-cranio-facial structures more accessible and the accurate examination of deep seated lesions possible. Orthodontic procedures require analysis of risk factors for root resorption [10,23,39], while stress distribution of orthodontic and restorative procedures require assessments of the root morphology and the long axis of the teeth [15,24,26]. Severe root resorption has been reported in 7% to 15% of the population, and in 73% of individuals who had orthodontic treatment [1,20]. Despite the progress, each imaging modality has its own limitations for diagnostic purposes and there is a clinical need to combine these imaging techniques for more comprehensive diagnostics and to assist treatment planning.

In this paper, we propose to automate the merging and annotation of CBCT with IOS imaging modalities. Both of these imaging techniques require machine learning methods to isolate anatomic structures of interest [5,9,17]. We describe our segmentation framework for both imaging modalities. Our CBCT image segmentation procedure outputs root canals [12] while IOS surface segmentation outputs individual crowns [11]. In our framework, we merge both outputs and label each individual root canal and dental crown following the 'Universal Numbering System' [13] that has been adopted by the American Dental Association (ADA) [2].

Our target application is the analysis of root canal and crown morphology and location. Specifically, the proposed clinical decision support application provides automatic visualization of anatomically complex root canal systems relative to accurate dental crowns for individually annotated teeth. The following sections describe the datasets, methods and results of this study.

2   Materials

The dataset consisted of 80 mandibular IOS (40 for the upper and 40 for the lower dental arches) and CBCT scans of the jaws for the same 80 subjects. The mandibular CBCT scans were obtained using the Vera-viewepocs 3D R100 (J Morita Corp.) with the following acquisition protocol: FOV 100 × 80 mm; 0.16 mm^3 voxel size; 90 kVp; 3 to 5 mA; and 9.3 s. Digital dental model of the mandibular arch was acquired from intra oral scanning with the TRIOS 3D intra oral scanner (3 Shape; software version: TRIOS 1.3.4.5). The TRIOS intra oral scanner (IOS) utilizes "ultrafast optical sectioning" and confocal microscopy to generate 3D images from multiple 2D images with an accuracy of 6.9 ± 0.9 mum. All scans were obtained according to the manufacturer's instructions, by one trained operator. Two open-source software packages, ITK-SNAP 3.8 [27] and Slicer 4.11 [7] were used to perform user interactive manual segmentation of the volumetric images and common spatial orientation of the mandibular dental arches to train our deep learning models. All IOS and CBCT scans were registered to each other using the validated protocol described by Ioshida *et al.* [14].

3 Related Work

3.1 Root Canal Segmentation Algorithm

Different approaches have been proposed to segment root canals from CBCT or micro-CT scans. Machado *et al.* [22] uses a semi-automated method to compute the minimum value for dentin tissue and applies a thresholding algorithm to extract the root canals. Sfeir *et al.* [32] enhances the scans using a super resolution technique and applies the Otsu's global thresholding algorithm. Other approaches use deep learning algorithms [9, 18, 40], particularly the U-Net architecture [31] developed for biomedical image segmentation. Our approach for CBCT root canal segmentation is similar to a previously reported method [18] where we extract 2D slices from our 3D volume and train a U-Net to segment the root canals.

3.2 3D Shape Analysis for Segmentation and Classification

There is a wide variety of approaches for shape analysis that can be broadly divided into two categories, *i.e.*, hand-crafted based methods and learning based methods. In general, hand-crafted methods describe shapes using fundamental solutions to equations [3, 6, 25, 29, 35], while learning-based methods use vast amounts of data to learn descriptors directly from the data [5, 9, 16, 19, 21, 28, 30, 36–38]. One of the main challenges in the learning-based methods category is to adapt state of the art algorithms for classification or segmentation to work on 3D shapes. The main impediment is the arbitrary structures of 3D models which are usually represented by point clouds or triangular meshes, whereas deep learning algorithms use the regular grid-like structures found in 2D/3D images. Multi-view methods offer an intuitive alternative to adapt these algorithms to work on 3D shapes. These approaches render a 3D object from different view-points and capture snapshots that can be used to extract 2D image features. Our approach for IOS segmentation falls in the multi-view category and is similar to previously reported work named FlyByCNN [5]. We use the multi-view approach for the segmentation task of dental crowns and a classification task of mandibles between upper and lower, we use the output of this task to label the dental crowns following the 'Universal Numbering System'.

4　Methods

4.1　Root Canal Segmentation Algorithm

Our pipeline for root canal segmentation is shown in Fig. 1. First we extract individual slices from the 3D CBCT scans. A total of 17,992 slices are used to train the U-Net. The upper left image in Fig. 1 shows a slice of the CBCT scan and the respective binary segmentation. An 8 folds cross-validation approach is used to tune the hyperparameters for training. The folds are created in a subject-wise manner, *i.e.*, all slices from a subject are in the same fold. The distribution of upper and lower CBCT mandibles are also balanced for training. A final model is trained using 64 scans (80%), the 16 remaining scans are used for testing (20%).

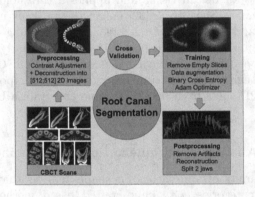

Fig. 1. Pipeline of the RCS algorithm

During training, we perform data augmentation by applying different transformations (Table 1) to the images (a heat map, that sums all the segmented scans and applied the random transformations, is shown in the upper right image in Fig. 1):

Table 1. Data augmentation transformations for the RCS algorithm

Random transformations	Rotation	Shift (x and y axes)	Shear	Zoom
Range values	0° to 360°	−30 to 30 pixels	−0.2 to 0.2% of the image	0 to 15 pixels

Training was done with 2 Nvidia Tesla 100 - 16 GB VRAM along with the following hyperparameters: Learning rate = 0.0008, Epoch = 100, Batch size = 16, Dropout = 0.1, Number of filters = 32.

4.2 Dental Model Segmentation Algorithm

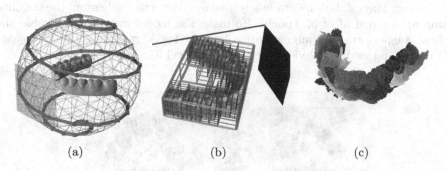

(a)   (b)   (c)

Fig. 2. a) Sphere sampling via icosahedron subdivision and spherical spiral. b) Octree subdivision surface sampling c) Each point in the 3D surface has a unique ID. The point ID is encoded in the RGB components and rendered.

Figure 2(a) shows 2 different types of sampling on a sphere where these vantage points are used to capture views of the 3D model. In our segmentation framework, we use the icosahedron subdivision to capture views of the 3D object. The IOS segmentation requires a mechanism to put back information into the 3D object after running inference on individual views. Our approach to generate the 1-1 relation of a 2D image and point in the mesh is done via our 'PointId map' generation algorithm. The PointId map makes it possible to extract properties from the mesh and/or to put information back into the mesh. Octrees are used to create a 1-1 relationship of pixels in the 2D image and triangles in the 3D object as previously reported in [5]. In this work we propose to generate 'PointId maps' by encoding each point in the mesh with a unique color. We use the following relationship to encode each unique point id into an RGB component

$$r = p_i \bmod 255.0 + 1$$
$$g = \lfloor p_i/255.0 \rfloor \bmod 255.0 \qquad (1)$$
$$b = \lfloor \lfloor p_i/255.0 \rfloor /255.0 \rfloor \bmod 255.0$$

where p_i is the unique id for a point in the mesh. Using our 'PointId maps' we achieve a 10 fold improvement in speed during inference compared to [5] as the algorithm uses the graphics processing unit (GPU) to link the 2D image view to points in the 3D mesh. In this work we also add an improvement in the post-processing stage of the IOS segmentation. The post-processing steps in [5] include island-removal (or removal of unlabeled points), labeling of each individual crown with a random number and erosion of the boundary between gum and teeth. We add an additional step which is the dilation of the boundary between gum and teeth after the island removal step. This prevents merging of some dental crowns where the boundary is not accurately detected (see Fig. 3 for the teeth segmentation steps). Training is done on a NVIDIA TITAN V - 12 GB

GPU. We use Categorical Crossentropy as a loss function with a learning rate of $1e^{-4}$. Our dataset is split by scans, 70% for training 10% for validation and 20% for testing. The validation data loss is used as criteria to 'early stop' the training. Training stopped after 32 epochs. To make the model more generalizable and robust, we perform data augmentation by applying 50 random rotations to each 3D model and extract 12 views (icosahedron level 0).

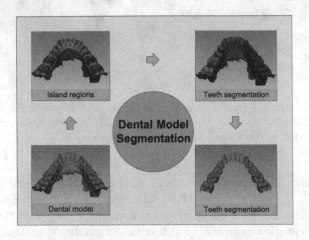

Fig. 3. Pipeline of the DMS algorithm

4.3   Universal Labeling and Merging Algorithm

The Universal Labeling and Merging (ULM) algorithm integrates the two imaging modalities and allows clinicians to save each individual tooth and its root canal in separate folders. Such output files allow clinicians to plan treatment for specific teeth or regions of the dental arch rather than a single whole dental arch surface mesh. The teeth were first labeled then the registered IOS and CBCT scans were automatically merged together. Each root inherited from the universal label of its corresponding tooth.

To label each tooth with the universal numbering system, it is necessary to know if the jaw is a lower or upper one. For this task, we trained a classification neural network that takes 2D views from the 3D object as inputs [5]. However, for this classification task we now use the spherical spiral sampling on the sphere and create a sequence of images (64 total). A pre-trained VGG19 [34] network initialized with weights trained on the ImageNet Dataset [8] is used to extract features from each view (this yields a matrix with shape [64, 512]). The image features are then fed to a bi-directional Long Short Term Memory (LSTM) layer with 64 units and a fully connected layer (64 units) followed by a softmax activation. The training was done on 37 scans (80% of the dataset). The 9 remaining scans (20% of the dataset) were used for testing. We performed data augmentation by applying 50 random rotations.

Once we know if it's an upper or lower jaw, the *Iterative Closest Point* algorithm is used to align the input surface to a template. The next step is to find the closest connected region label to assign the universal labels of the chosen template.

The pipeline of our project is shown in Fig. 4. ULM takes the outputs of the 2 previous segmentations, assigns universal labels on each tooth and merges them into a single model. The output of the pipeline can be used for clinical decision support purposes.

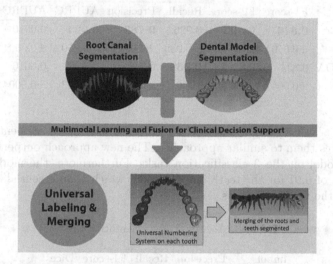

Fig. 4. Pipeline of the ULM algorithm

5   Results

5.1   Root Canal Segmentation

The pixels that belong to the segmentation represent a small proportion of the whole scan (0.1%) compared to the background. To ensure that there is no class imbalance between those pixels, the Area Under the Precision Recall Curve (AUPRC) metric was computed. First the AUPRC baseline is calculated by dividing the number of pixels belonging to the root canal by the total number of pixels in the scan. Then, without any post processing steps, thresholds (from 1 to 255) are applied on the scan to calculate the area under the curve. This metric shows that if the AUPRC is above the baseline then the model is able to learn even in the case of a class imbalance. Table 2 demonstrates that the RCS model is able to learn properly.

Other metrics such as the recall, precision, F1-score and F2-score were also computed to know how efficient the model is. Two model were compared to each other as well, one with data augmentation and the other without it. As shown on Table 2, the model with data augmentation outperformed the model without DA for all metrics except for Recall. Data augmentation also helped increase the robustness as it decreaseed the standard deviation (std). From a clinical point of view, it is better to have the root canal too much segmented than no root canal. That's why the F2-score was computed as it weights twice more to compared to precision.

Table 2. Metrics for the root canal segmentation model

Model	F1-score	F2-score	Recall	Precision	AUPRC	AUPRC baseline
With DA	**0.846**	**0.892**	0.926	**0.782**	**0.924**	0.0010
Std	±0.050	±0.038	±0.039	±0.074	±0.050	±0.0004
Without DA	0.834	0.889	**0.932**	0.763	0.858	0.0010
Std	±0.065	±0.044	±0.046	±0.102	±0.113	±0.0004

Table 3 as well as Fig. 5 shows the results for the image segmentation task and compares them to similar approaches. The new approach outperformed the previous model on the 3 classification tasks. Furthermore, the model reached an F1-score of 0.91 for the teeth classification, which is an essential step in the merging of the teeth and roots.

Table 3. Accuracy of segmentation task

Label	Precision	Recall	F1-score	Dice
Boumbolo et al. [5]				
Gum	0.92	0.76	0.83	-
Teeth	0.87	0.77	0.82	-
Boundary	0.57	0.86	0.69	-
Dental model segmentation algorithm				
Gum	0.98	0.97	**0.97**	0.97
Teeth	0.96	0.86	**0.91**	0.90
Boundary	0.82	0.93	**0.87**	0.87

5.2   Universal Labeling and Merging Algorithm

The results of the mandible classification are shown in the first part of the Table 4. We obtained an F1-score of 0.9. The Universal Labeling and Merging algorithm results are in the second part of Table 4, and the model achieved an F1-score of 0.85. The training of the automatic dental crown segmentation algorithm was done with digital dental models of permanent full dentition. Preliminary testing of the trained model includes segmentation of dental crowns of the digital dental models in the primary and mixed stages of the dentition, as well as cases with unerupted, missing or ectopically positioned teeth.

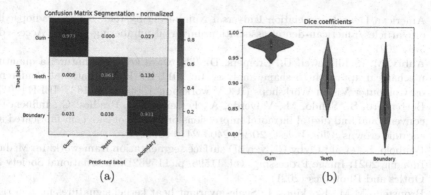

Fig. 5. a) Results of the teeth, gum and boundary prediction b) Dice coefficient of the teeth, gum and boundary

Table 4. Accuracy of the classification mandibular and the universal labels

	Precision	Recall	F1-score	Testing dataset
Mandibular classification				
Accuracy	*NA*	*NA*	**0.90**	80
Macro avg	0.92	0.90	0.90	80
Weight avg	0.92	0.90	0.90	80
Universal labels classification				
Accuracy	*NA*	*NA*	**0.85**	3647876
Macro avg	0.85	0.82	0.82	3647876
Weight avg	0.86	0.85	0.84	3647876

6   Conclusion

Merging and annotation of CBCT and IOS imaging modalities will help guide clinical decision support and quantitative treatment planning for specific teeth, implant placement, root canal treatment, restorative procedures, and biomechanics of tooth movement in orthodontics. The automatic identification of each tooth accurate crown and root canal morphology provides visualization the anatomically complex root canal systems, obliterated canals, and aids treatment planning for image-guided root canal treatment, regenerative endodontic therapies, position of implant placement, restorative crown shape planning to avoid inadequate forces on roots and planning of orthodontic tooth movement.

References

1. Ahlbrecht, C.A., et al.: Three-dimensional characterization of root morphology for maxillary incisors. PLoS One **12**(6), e0178728 (2017). https://doi.org/10.1371/journal.pone.0178728

2. American Dental Association Universal Numbering System. https://radiopaedia. org/articles/american-dental-association-universal-numbering-system. Accessed 5 July 2021

3. Aubry, M., Schlickewei, U., Cremers, D.: The wave kernel signature: a quantum mechanical approach to shape analysis. In: 2011 IEEE International Conference on Computer Vision Workshops (ICCV workshops), pp. 1626–1633. IEEE (2011)

4. Berrendero, S., Salido, M., Valverde, A., Ferreiroa, A., Pradíes, G.: Influence of conventional and digital intraoral impressions on the fit of cad/cam-fabricated all-ceramic crowns. Clin. Invest. **20**(9), 2403–2410 (2016)

5. Boubolo, L., et al.: Flyby CNN: a 3D surface segmentation framework. In: Medical Imaging 2021: Image Processing. vol. 11596, p. 115962B. International Society for Optics and Photonics (2021)

6. Bronstein, M.M., Kokkinos, I.: Scale-invariant heat kernel signatures for non-rigid shape recognition. In: 2010 IEEE Computer Society Conference on Computer Vision and Pattern Recognition, pp. 1704–1711. IEEE (2010)

7. Citing slicer. https://www.slicer.org/w/index.php?title=CitingSlicer&oldid=63 090

8. Deng, J., Dong, W., Socher, R., Li, L.J., Li, K., Fei-Fei, L.: Imagenet: a large-scale hierarchical image database. In: 2009 IEEE Conference on Computer Vision and Pattern Recognition, pp. 248–255. IEEE (2009)

9. Dumont, M., et al.: Patient specific classification of dental root canal and crown shape. In: International Workshop on Shape in Medical Imaging, pp. 145–153. Springer (2020)

10. Elhaddaoui, R, et al.: Resorption of maxillary incisors after orthodontic treatment-clinical study of risk factors. Int. Orthod. **14**, 48–64. (2016). https://doi.org/10. 1016/j.ortho.2015.12.015

11. Dental model segmentation. https://github.com/DCBIA-OrthoLab/fly-by-cnn

12. Root canal segmentation. https://github.com/RomainUSA/CBCT_seg

13. Universal labeling and merging. https://github.com/DCBIA-OrthoLab/fly-by-cnn/blob/master/src/py/labeling.py, https://github.com/DCBIA-OrthoLab/fly-by-cnn/tree/master/src/py/PSCP

14. Ioshida, M, et al.: Accuracy and reliability of mandibular digital model registration with use of the mucogingival junction as the reference. Oral Surg. Oral Med. Oral Pathol. Oral Radiol. **127**(4), 351–360 (2018). https://doi.org/10.1016/j.oooo.2018. 10.003

15. Kamble, R.H., et al.: Stress distribution pattern in a root of maxillary central incisor having various root morphologies: a finite element study. Angle Orthod. **82**, 799–805 (2012). https://doi.org/10.2319/083111-560.1

16. Kanezaki, A., et al.: Rotationnet: joint object categorization and pose estimation using multiviews from unsupervised viewpoints. In: Proceedings of the IEEE Conference on Computer Vision and Pattern Recognition, pp. 5010–5019 (2018)

17. Ko, C.C., et al.: Machine learning in orthodontics: application review. Craniof. Growth Ser. **56**, 117–135 (2020). http://hdl.handle.net/2027.42/153991

18. Kwak, G.H., et al.: Automatic mandibular canal detection using a deep convolutional neural network. Sci. Rep. **10**, 5711 (2020). https://doi.org/10.1038/s41598-020-62586-8

19. Liu, M., Yao, F., Choi, C., Sinha, A., Ramani, K.: Deep learning 3D shapes using alt-az anisotropic 2-sphere convolution. In: International Conference on Learning Representations (2018)

20. Lupi, J.E., et al.: Prevalence and severity of apical root resorption and alveolar bone loss in orthodontically treated adults. Am. J. Orthod. Dentofacial. Orthop. **109**(1), 28–37 (1996). https://doi.org/10.1016/s0889-5406(96)70160-9
21. Ma, C., Guo, Y., Yang, J., An, W.: Learning multi-view representation with ISTM for 3-D shape recognition and retrieval. IEEE Trans. Multimedia **21**(5), 1169–1182 (2018)
22. Machado, J., Pires, P., Santos, T., Neves, A., Lopes, R., Visconti, M.A.: Root canal segmentation in cone-beam computed tomography. Braz. J. Oral Sci. **18**, e191627 (2019). DOIurl10.20396/bjos.v18i0.8657328
23. Marques, L.S., et al.: Severe root resorption in orthodontic patients treated with the edgewise method: prevalence and predictive factors. Am. J. Orthod. Dentofacial. Orthop. **137**, 384–388 (2010) https://doi.org/10.1016/j.ajodo.2008.04.024
24. Marques, L.S., et al.: Severe root resorption and orthodontic treatment: clinical implications after 25 years of follow-up. Am. J. Orthod. Dentofac. Orthop. **139**, S166–5169 (2011). https://doi.org/10.1016/j.ajodo.2009.05.032
25. Ming-Kuei, H.: Visual pattern recognition by moment invariants. IRE Trans. Inf. Theory **8**(2), 179–187 (1962)
26. Oyama, K., et al.: Effects of root morphology on stress distribution at the root apex. Eur. J. Orthod. **29**, 113–117 (2007). https://doi.org/10.1093/ejo/cjl043
27. Paul, A., et al.: User-guided 3D active contour segmentation of anatomical structures: Significantly improved efficiency and reliability. Neuroimage **31**(3), 1116–1128 (2006). https://www.itksnap.org
28. Qi, C.R., Su, H., Mo, K., Guibas, L.J.: Pointnet: deep learning on point sets for 3D classification and segmentation. In: Proceedings of the IEEE Conference on Computer Vision and Pattern Recognition, pp. 652–660 (2017)
29. Ribera, N., et al.: Shape variation analyzer: a classifier for temporomandibular joint damaged by osteoarthritis. In: Medical Imaging 2019: Computer-Aided Diagnosis, vol. 10950, p. 1095021. International Society for Optics and Photonics (2019)
30. Riegler, G., Osman Ulusoy, A., Geiger, A.: Octnet: learning deep 3D representations at high resolutions. In: Proceedings of the IEEE Conference on Computer Vision and Pattern Recognition, pp. 3577–3586 (2017)
31. Ronneberger, O., Fischer, P., Brox, T.: U-Net: convolutional networks for biomedical image segmentation. In: Navab, N., Hornegger, J., Wells, W.M., Frangi, A.F. (eds.) MICCAI 2015. LNCS, vol. 9351, pp. 234–241. Springer, Cham (2015). https://doi.org/10.1007/978-3-319-24574-4_28
32. Sfeir, R., Michetti, J., Chebaro, B., Diemer, F., Basarab, A., Kouamé, D.: Dental root canal segmentation from super-resolved 3D cone beam computed tomography data. In: 2017 IEEE Nuclear Science Symposium and Medical Imaging Conference (NSS/MIC), pp. 1–2 (2017). https://doi.org/10.1109/NSSMIC.2017.8533054
33. Shah, N., Bansal, N., Logani, A.: Recent advances in imaging technologies in dentistry. World J. Radiol. **6**(10), 794 (2014)
34. Simonyan, K., Zisserman, A.: Very deep convolutional networks for large-scale image recognition. arXiv preprint arXiv:1409.1556 (2014)
35. Styner, M., et al.: Framework for the statistical shape analysis of brain structures using SPHARM-PDM. Insight J. **2006**(1071), 242 (2006)
36. Su, H., Maji, S., Kalogerakis, E., Learned-Miller, E.: Multi-view convolutional neural networks for 3D shape recognition. In: Proceedings of the IEEE International Conference on Computer Vision, pp. 945–953 (2015)
37. Wang, P.S., Liu, Y., Guo, Y.X., Sun, C.Y., Tong, X.: O-cnn: Octree-based convolutional neural networks for 3d shape analysis. ACM Transactions on Graphics (TOG) **36**(4), 1–11 (2017)

38. Wu, Z., et al.: 3D shapenets: a deep representation for volumetric shapes. In: Proceedings of the IEEE Conference on Computer Vision and Pattern Recognition, pp. 1912–1920 (2015)
39. Xu, X., et al.: 3D tooth segmentation and labeling using deep convolutional neural networks. IEEE Trans. Vis. Comput. Graph. 2019 **25**(7), 2336–2348. (2019) https://doi.org/10.1109/TVCG.2018.2839685
40. Zichun, Y., et al.: CBCT image segmentation of tooth-root canal based on improved level set algorithm. In: Proceedings of the 2020 International Conference on Computers, Information Processing and Advanced Education (CIPAE 2020), pp. 42–51. Association for Computing Machinery, New York (2020). https://doi.org/10.1145/3419635.3419654

Structure and Feature Based Graph U-Net for Early Alzheimer's Disease Prediction

Yun Zhu, Xuegang Song, Yali Qiu, Chen Zhao, and Baiying Lei[✉]

National-Region Key Technology Engineering Laboratory for Medical Ultrasound, Guangdong Key Laboratory for Biomedical Measurements and Ultrasound Imaging, School of Biomedical Engineering, Health Science Center, Shenzhen University, Shenzhen 518060, China
leiby@szu.edu.cn

Abstract. Alzheimer's disease (AD) is a common neurodegenerative brain disease, which seriously affects the quality of life. Predicting its early stage (e.g., mild cognitive impairment (MCI) and significant memory concern (SMC)) has great significance for early diagnosis. As the vague imaging features of MCI and SMC, graph convolution network (GCN) has been widely used as its advantage of fusing phenotypic information (e.g., gender and age) and establishing relationship between subjects for filtering. Graph U-Net can integrate GCN into U-Net structure with promising classification performance, but it ignores the structure information of graph in the pooling process and leads to the loss of important nodes. To capture the high-order information in the graph, and integrate the structure and node feature information in its pooling operation, a structure and feature based graph U-Net (SFG U-Net) is proposed to predict MCI and SMC in this paper. Firstly, we use the sliding window method to construct dynamic functional connection network (FCN) based on functional magnetic resonance imaging (fMRI). Secondly, we combine image information and phenotypic information to construct functional graph. Thirdly, the structure and the feature of the node in graph are considered in the adaptive pooling layer. Lastly, we get the final diagnosis result by inputting the graph into SFG U-Net. The proposed method is validated on the public data set of the Alzheimer's Disease Neuroimaging Initiative (ADNI), which achieves a mean classification accuracy of 83.69%.

Keywords: Mild cognitive impairment · Significant memory concern · Functional magnetic resonance imaging · Structure and feature based graph U-Net

1 Introduction

Alzheimer's disease (AD) is a neurodegenerative brain disease, and there is no effective medicine to cure it [1]. Mild cognitive impairment (MCI) and significant memory concern (SMC) are the early stage of AD, which can be effectively delayed or prevented the conversion to AD with timely treatment [2, 3]. Studies have shown that MCI has an annual 10–15% rate to get worse and more than 50% rate within five years to convert

T. Syeda-Mahmood et al. (Eds.): ML-CDS 2021, LNCS 13050, pp. 93–104, 2021.
https://doi.org/10.1007/978-3-030-89847-2_9

to AD [4]. Due to the high conversion rate of MCI to AD, the timely diagnosis and classification of MCI has extracted wide attention [5–8].

In recent years, the development of neuroimaging has provided effective detection methods for MCI and SMC. Functional magnetic resonance imaging (fMRI) [9], which utilizes blood oxygen-dependent level signals to perform real-time living imaging of brain with the characteristics of non-radioactive and high spatial resolution, has been widely used for brain functional connection network (FCN) construction[10]. By extracting biomarkers from FCN and then using machine learning, the disease can be diagnosed successfully [11–13]. In the construction of brain functional network, PCC is intuitive and easy to interpret. Most PCC based methods focus on constructing traditional static brain FCN. However, growing evidence suggests that brain FCN is actually varying across time and such variation could not only derive from noise. Then, Schwab *et al.* [14] proposed a dynamic brain FCN construction method and obtained better results than static brain networks. Therefore, we realize the diagnosis task based on dynamic brain FCN in this paper.

As the insignificant imaging features of MCI and SMC, their diagnosis performance is usually unsatisfied. Then phenotypic information (e.g., gender and age) is proposed as a supplement to improve diagnosis performance. Recently, graph convolution network (GCN) has been widely used to fuse phenotypic information (e.g., gender and age) and establish relationship between subjects for filtering. Experimental results show that the mean classification accuracy has improved to 86.83% in the work [15]. However, a single GCN cannot extract the high-order information of the graph. To solve this problem, many methods are proposed [16–18]. Based on the good performance of U-Net [19] in image classification and segmentation, Gao and Ji *et al.* [16] proposed a model named Graph U-Net by using GCN to replace the convolution block and design a novel pooling layer. The experimental results show that their model achieves good performance. Furthermore, Lostar *et al.* [17] used Hypergraph U-Net structure for brain graph embedding and classification, and also used GCN to replace the original convolution. Similarly, their results show that the method is effective. The Pooling layer in above works plays an extremely important role, which is used to extract high-order information and reduce the size of the graph. The top-k graph pooling method in Graph U-Net [16] uses the node characteristics of the graph to select the first k important nodes to construct a new graph. It ignores the topological structure information of the graph and loses some important nodes in the down sampling process, which leads to poor detection effect. To solve this problem, an adaptive GCN and an adaptive fully connected layers (FC) are used to form the pooling layer in this paper. Based on this, we can combine the result from graph topology information and node feature information.

First of all, we use the sliding window and PCC to construct a dynamic FCN. Secondly, we construct the graph to combine image and phenotypic information. Thirdly, SFG U-Net is proposed for final diagnosis, where an adaptive GCN and an adaptive fully connected layers (FC) are used to form the pooling layer. The proposed method is verified on the ADNI dataset, and the results show that our method achieves promising result for predicting MCI and SMC. Generally, we propose a novel pooling layer in graph U-Net to capture the high-order information by considering the node feature information and graph topology information.

2 Methods

The model proposed in this paper is mainly composed of the following three parts.1) Brain network construction. The dynamic FCN based on fMRI data is constructed for each subject. For dynamic FCN, we use the sliding window to segment fMRI time series into multiple sub-series, and then use PCC to construct FCN between each ROI pairs. 2) Graph construction. We construct a functional graph based on above image information and their phenotypic information. Specifically, the edge connections of a graph are constructed based on phenotypic information, and their edge weights are computed by using PCC method to simulate their similarity between image information. 3) Classification. By inputting the above functional graph into the proposed SFG U-Net, we can get the final classification results. The overview of the proposed model is shown in Fig. 1.

2.1 Dynamic FCN

We preprocess fMRI data before constructing brain network. Firstly, we use head motion correction to remove motion artifacts in the fMRI time series, and use a Gaussian kernel with a maximum of 4mm to remove low-frequency noise and high-frequency noise. Then the time series of 90 ROIs of each subject was obtained through automatic anatomical labeling (AAL). After the above preprocessing, a sliding window is used to divide the entire fMRI time series into multiple sub-sequences. Then, the PCC is used to calculate the correlation of the sub-sequences of different ROIs. The specific construction method is as follows:

Assuming that the time series of fMRI has a total of W time points, the number of sub-sequences T is shown in the following formula:

$$T = [(W - L)/h] + 1, \tag{1}$$

where L represents the size of the sliding window, and h represents the sliding step. Then the tth sub-sequence can be expressed as:
$\mathbf{X}^{(t)} = \left[\mathbf{x}_1^{(t)}, \mathbf{x}_2^{(t)}, \ldots, \mathbf{x}_i^{(t)}, \ldots, \mathbf{x}_N^{(t)} \right] \in \mathbb{R}^{1 \times N}, t = 1, 2, \ldots, T$, where N represents the number of subjects, and the subject i can be represented by $\mathbf{x}_i^{(t)} = [ROI(1), ROI(2), \ldots, ROI(R)] \in \mathbb{R}^{R \times L}$, where R represents the number of ROIs, $R = 90$ in this paper. Based on $\mathbf{x}_i$, the dynamic FCN of subject i (represented by F_i) is constructed, which includes a series of sub-network $F_i^{(t)}$. The $F_i^{(t)}$ is constructed based on corresponding sub-sequence $\mathbf{x}_i^{(t)}$, which is defined as follows:

$$F_i^{(t)} = corr\{ROI(m), ROI(n)\}. \tag{2}$$

The dynamic FCN of the entire time series is:

$$C = \sum\nolimits_{t-1}^{T} \{F_1^{(t)}, F_2^{(t)}, \ldots, F_i^{(t)}, \ldots, F_N^{(t)}\}/T \in \mathbb{R}^{N \times 90 \times 90} \tag{3}$$

After constructing the dynamic FCN, each subject gets a 90×90 brain dynamic FCN.

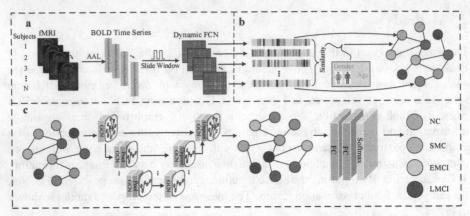

Fig. 1. The overall framework of our proposed model. **a)** Dynamic FCN. We utilize the fMRI data to construct a dynamic FCN for each subject. The dimension of the dynamic FCN for each subject is 90 × 90. **b)** Graph construction. By using feature extraction, we get a functional feature vector for each subject, and use them represent the node on graph. The edges between nodes are constructed by computing the similarity between the feature vectors and phenotypic information. Finally, we construct a sparse functional graph for N subjects. **c)** SFG U-Net. By inputting above graphs into SFG U-Net, we get a new graph for N subjects, and inputting the node features of the new graph into FC for final classification.

2.2   Graph Construction

After constructing the brain network, we obtain a 90 × 90 FCN for each subject. By extracting their upper triangular elements, we can get a 1 × 4005 function feature vector x_i^f for subject i. The image information and phenotypic information is used to construct the graph in this paper. $G(V, E, A)$ represents the undirected sparse graph. V includes all nodes on a graph and is represented by their feature vectors, where $V \in \mathbb{R}^{N \times 4005}$. E represents the edges between nodes. Edge connection is established based on phenotypic information, and edge weights are constructed by computing the similarity between the feature vectors. Then adjacency matrix A is constructed by combining edge connections and edge weights. Specifically, the function adjacency matrix A_{ij}^f is represented as:

$$A_{ij}^f = S\left(x_i^f, x_j^f\right) \times \left(P(B_i, B_j) + P(G_i, G_j)\right), \tag{4}$$

where x_i^f and x_j^f represent the functional feature vectors of subject i and subject j. B_i and B_j represent the gender information of subjects i and j, G_i and G_j represent the age information of subjects i and j. S represents the similarity between feature vectors, and the calculation method is defined as follows:

$$S\left(x_i^f, x_j^f\right) = \frac{\left(x_i^f - \overline{x}_i^f\right) \odot \left(x_j^f - \overline{x}_j^f\right)}{\left\| \left(x_i^f - \overline{x}_i^f\right) \right\|_2 \times \left\| \left(x_j^f - \overline{x}_j^f\right) \right\|_2}, \tag{5}$$

where $\bar{x}_i^f$ represents the average value of x_i^f, $\bar{x}_j^f$ represents the average value of x_j^f, the symbol $\odot$ represents dot product, and $P(\cdot)$ represents the relationship between phenotypic information, as shown in followings:

$$P(B_i, B_j) = \begin{cases} 1, B_i = B_j \\ 0, B_i \neq B_j \end{cases}, \quad P(G_i, G_j) = \begin{cases} 1, |G_i - G_j| < 3 \\ 0, else \end{cases}. \tag{6}$$

2.3 SFG U-Net

Since encoding and decoding structure of U-Net has achieved promising results in pixel-level prediction tasks, we construct a U-Net structure for classification. In our proposed SFG U-Net structure, we first use a Chebyshev GCN to perform a feature extraction on the input graph to reduce node feature dimension. Our encoding blocks are composed of GCN and Pool, where Pool extracts the main graph nodes to encode their high-order features, and GCN gathers the first-order information of each node. Decoding part is composed of GCN and unpool, where unpool restores the number of nodes to initial state, and GCN gathers information from neighbors.

2.3.1 Graph Convolution

The graph convolution operation in Kipf et al. [20] has been validated to be useful in classification tasks, which multiply the signal $h \in \mathbb{R}^N$ (the scalar of each node) with the filter $g_\gamma = diag(\gamma)$ (represented by the parameter $\gamma \in \mathbb{R}^N$ in the Fourier domain). It can be represented as follows:

$$g_\gamma * h = U g_\gamma U' h, \tag{7}$$

where U is the eigenvector matrix of the normalized graph Laplacian $L = I_N - D^{-\frac{1}{2}} A D^{-\frac{1}{2}} = U \wedge U'$, and A is the adjacency matrix of the graph, $\wedge$ is the diagonal matrix of eigenvalues, and $U' h$ is the Fourier transform of h, which transforms convolution of time domain into multiplication of frequency domain. Chebyshev polynomials is denoted as: $T_k(h) = 2h T_{k-1}(h) - T_{k-2}(h)$, $T_0(h) = 1$, $T_1(h) = h$.

By applying Chebyshev polynomials and using other simplification methods mentioned in Kipf et al. [20], Eq. (7) is reformulated as:

$$g_\gamma * h \approx \gamma \left(I_N + D^{-\frac{1}{2}} A D^{-\frac{1}{2}} \right) h, \tag{8}$$

To strengthen the self-characteristics of the nodes in the graph, we can set $\tilde{A} = A + I_N$, then $I_N + D^{-\frac{1}{2}} A D^{-\frac{1}{2}} \rightarrow \tilde{D}^{-\frac{1}{2}} \tilde{A} \tilde{D}^{-\frac{1}{2}}$, where $\tilde{D}_{ii} = \sum_j \tilde{A}_{ij}$.

Therefore, the convolution of a signal $H \in \mathbb{R}^{N \times C}$ (e.g., the C dimensional feature vector of each node) and the filter g_γ can be simplified as follows:

$$Z = \tilde{D}^{-\frac{1}{2}} \tilde{A} \tilde{D}^{-\frac{1}{2}} H \Theta, \tag{9}$$

where $\Theta \in \mathbb{R}^{C \times F}$ is the matrix parameter of the filter, and $Z \in \mathbb{R}^{N \wedge F}$ is the convolution signal matrix.

2.3.2 Pool Layer

As shown in Fig. 2, we build a FC, which takes the feature information of the nodes into account, and obtains the scores y_1^l from the node features. We also construct a GCN, which considers the topological structure information of our graph, and gets the scores y_2^l from the topological structure information. Finally, we set a parameter a to adaptively adjust the importance of the information to select the best node. Assuming there are N nodes in this paper, the feature matrix of all input nodes is represented as $V^l \in \mathbb{R}^{N \times C}$, and the adjacency matrix of the input graph is represented as $A^l \in \mathbb{R}^{N \times N}$. Then the output of the FC corresponding to the feature matrix of the input node is $y_1^l \in \mathbb{R}^{N \times 1}$, and the output of the GCN corresponding to the adjacency matrix of the input graph and the feature of the graph node is $y_2^l \in \mathbb{R}^{N \times 1}$. Performing adaptive weighted summation according to y_1^l and y_2^l to get $y^l \in \mathbb{R}^{N \times 1}$, and then select the best node according to y^l to construct a new graph. The specific expression is as follows:

$$
\begin{aligned}
y_1^l &= FC(V^l), \\
y_2^l &= GCN(V^l), \\
y^l &= a \times y_1^l + (1 - a) \times y_2^l, \\
idx &= Top_{k(y^l)}, \\
V^{l+1} &= V^l(idx, :), \\
A^{l+1} &= A^l(idx, idx),
\end{aligned}
\tag{10}
$$

where idx represents the position information of the first K nodes with relatively large scores in y^l, so that suitable nodes can be selected to construct a new graph. It is worth noting that the value a can be changed to select the best node. In addition, the parameters of the FC and GCN in this paper can be optimized during the back propagation process, so that the accuracy of the final output result is improved.

2.3.3 Unpool Layer and Skip Connection

For the decoding part of the graph data, we construct the same number of decoding blocks as the encoding part, and each decoding block is composed of GCN and unpool layer. Unpool layer performs the inverse operation of pool and restores the graph data to its original input structure. Specifically, we first record the position index value of the selected node corresponding to each pool layer, and use these position index values to restore the graph data of the unpool layer corresponding to the decoding block. Secondly, due to the positions of some nodes in the graph obtained by unpool layer are null values, the node information is filled by skip connections. Finally, we use the restored graph data as the input of the next decoding block, and then restore the graph data.

The skip connection used in this paper is an add operation. The add operation is performed on the corresponding encoding block and decoding block to enhance the graph data information, combining high-order and low-order information. We add the initial input graph data to the output graph data of the corresponding last layer of decoding structure, and use this data for the final prediction task.

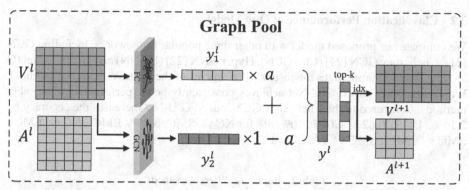

Fig. 2. The framework of the proposed Graph Pool. Supposing that the number of nodes in the input graph of this layer is 6, the node feature is 8, and the k in the top-k is set to 4. The FC evaluates the node feature information of the graph, and outputs a 6×1 vector y_1^l. Meanwhile, GCN is also used to evaluate the topological structure information of the graph, and a 6×1 vector y_2^l is output by combining the topological structure and the node feature information of the graph. Finally, we adaptively combine the output vectors of the two to obtain a 6×1 vector y^l to determine the position index values of the first 4 nodes. Using these index values to obtain a 4×8 node feature matrix V^{l+1} of the next layer and a 4×4 adjacency matrix A^{l+1}.

3   Experiments and Results

In this paper, we evaluate our proposed framework on the ADNI dataset. All experiments are performed by using 10-fold cross-validation strategy. There are three types of diseases that include SMC, early MCI (EMCI) and later MCI (LMCI). Our classification tasks are NC vs. SMC, NC vs. EMCI, NC vs. LMCI, SMC vs. EMCI, SMC vs. LMCI and EMCI vs. LMCI. The model parameters of this paper are set as follows: learning rate is 0.01, GCN dropout is 0.3, FC dropout is 0.2, hidden layer is 48, the layers of SFG U-Net is 3. The evaluation indicators used in this experiment are accuracy (ACC), sensitivity (SEN), specificity (SPEC), F score (F1-score) and area under the surface (AUC).

3.1   Dataset

We collected a dataset of 291 subjects from ADNI, which includes SMC, EMCI, LMCI and NC. The dataset includes the fMRI data and the phenotypic information (e.g. gender and age). The details of this information are shown in Table 1.

Table 1. Summary of the data.

Group	SMC(44)	EMCI(86)	LMCI(67)	NC(94)
Male/Female	17M/27F	43M/43F	36M/31F	42M/53F
Age (mean ± SD)	76.31 ± 5.41	75.09 ± 6.44	74.31 ±7.93	75.57 ± 5.98

3.2 Classification Performance of Our Model

We compare our proposed model with other three popular frameworks, including GCN [21], Chebyshev GCN [21] (Che-GCN), Hyper GNN [22] (HGNN) and Graph U-Net (G U-Net). Table 2 summarizes the results. ROC curves of the six tasks is shown in Fig. 3. We can find that our SFG U-Net achieves consistently better performance than other methods. Compared to the Chebyshev GCN, our SFG U-Net increases the accuracy by 2.14%, 4.44%, 0.62%, 4.68%, 10%, 8% for NC vs. SMC, NC vs. EMCI, NC vs. LMCI, SMC vs. EMCI, SMC vs. LMCI and EMCI vs. LMCI.

Table 2. Results of different methods.

Method	NC vs. SMC					NC vs. EMCI				
	ACC	SEN	SPEC	F1	AUC	ACC	SEN	SPEC	F1	AUC
HGNN	0.7252	0.0263	0.99	0.0500	0.5081	0.5444	0.4137	0.6666	0.4675	0.5402
GCN	0.7874	0.5714	0.8842	0.6234	0.7278	0.7611	0.7093	0.8085	0.7394	0.7589
Che-GCN	0.8110	0.4523	**0.9684**	0.5937	0.7104	0.7556	**0.7906**	0.7234	0.7555	0.7570
G U-Net	0.8033	0.5349	0.9255	0.6301	0.7302	0.7833	0.5698	**0.9787**	0.7153	0.7742
SFG U-Net	**0.8324**	**0.6429**	0.9158	**0.7013**	**0.7793**	**0.8000**	0.7674	0.8298	**0.7857**	**0.7986**

Method	NC vs. LMCI					SMC vs. EMCI				
	ACC	SEN	SPEC	F1	AUC	ACC	SEN	SPEC	F1	AUC
HGNN	0.5897	0.1714	0.9121	0.2667	0.5418	0.6846	0.9438	0.1219	0.8038	0.5329
GCN	0.7651	0.6119	0.8723	0.6833	0.7421	0.7449	0.8139	0.6046	0.8092	0.7093
Che-GCN	0.8015	**0.7761**	0.8191	0.7647	**0.7976**	0.8147	0.8953	**0.6511**	0.8651	0.7732
G U-Net	0.8026	0.5224	**1.0000**	0.6863	0.7612	0.8000	0.9767	0.4418	0.8660	0.7093
SFG U-Net	**0.8077**	0.7165	0.8723	**0.7560**	0.7944	**0.8615**	0.9767	0.6279	**0.9032**	**0.8023**

Method	SMC vs. LMCI					EMCI vs. LMCI				
	ACC	SEN	SPEC	F1	AUC	ACC	SEN	SPEC	F1	AUC
HGNN	0.6846	0.9438	0.1219	0.8038	0.5328	0.5096	0.1733	0.8333	0.2574	0.5033
GCN	0.7273	0.8805	0.4883	0.7973	0.6845	0.8117	**0.8030**	0.8161	0.7852	0.8096
Che-GCN	0.7818	0.8955	0.6047	0.8333	0.7501	0.7579	0.5909	0.8850	0.6782	0.7379
G U-Net	0.7727	0.9403	0.5116	0.8344	0.7259	0.7954	0.6268	**0.9186**	0.7241	0.7302
SFG U-Net	**0.8818**	**0.9701**	**0.7442**	**0.9091**	**0.8571**	**0.8379**	0.7576	0.8966	**0.8000**	**0.8271**

3.3 The Effect of Our Pool Layer

To validate the effectiveness of our pool layer that considers both graph structure information and node feature information. We conduct experiments by removing FC (removing the information of node feature) or GCN (removing the information of graph structure) for comparison with our SFG U-Net. The results are summarized in Table 3. In Table 3, the FG U-Net represent the model that only considering the information of node feature in the pool layer, the SG U-Net represented the model only considering the information of the graph structure, and the SFG U-Net represent the model considering both above two information in the pool layer. Experimental results show that our pool layer achieves better performance in all of the six tasks and our pool layer can maximum increase 8.18% accuracy for SMC vs. LMCI.

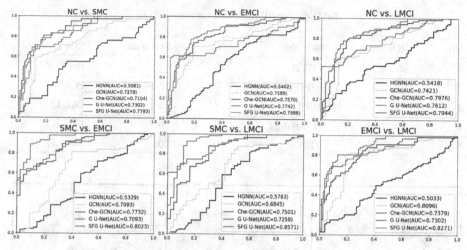

Fig. 3. ROC curves of different methods.

Table 3. Results of pool layer by removing FC or GCN.

Method	NC vs. SMC					NC vs. EMCI				
	ACC	SEN	SPEC	F1	AUC	ACC	SEN	SPEC	F1	AUC
FG U-Net	0.8033	0.5349	0.9255	0.6301	0.7302	0.7833	**0.5698**	0.9787	0.7153	0.7742
SG U-Net	0.8104	0.5476	**0.9263**	0.6389	0.7370	0.7778	0.7674	0.7872	0.7674	0.7773
SFG U-Net	**0.8324**	**0.6429**	0.9158	**0.7013**	**0.7793**	**0.8000**	0.7674	**0.8298**	**0.7857**	**0.7986**

Method	NC vs. LMCI					SMC vs. EMCI				
	ACC	SEN	SPEC	F1	AUC	ACC	SEN	SPEC	F1	AUC
FG U-Net	0.8026	0.5224	**1.0000**	0.6863	0.7612	0.8000	0.9767	0.4418	0.8660	0.7093
SG U-Net	0.7644	0.5970	0.830	0.6780	0.7400	0.8057	**0.9883**	0.4419	0.8718	0.7151
SFG U-Net	**0.8077**	**0.7165**	0.8723	**0.7560**	**0.7944**	**0.8615**	0.9767	**0.6279**	**0.9032**	**0.8023**

Method	SMC vs. LMCI					EMCI vs. LMCI				
	ACC	SEN	SPEC	F1	AUC	ACC	SEN	SPEC	F1	AUC
FG U-Net	0.7727	0.9403	0.5116	0.8344	0.7259	0.7954	0.6268	**0.9186**	0.7241	0.7302
SG U-Net	0.8000	**1.0000**	0.4884	0.8589	0.7441	0.8317	0.7273	0.9080	0.7869	0.8177
SFG U-Net	**0.8818**	0.9701	**0.7442**	**0.9091**	**0.8571**	**0.8379**	0.7576	0.8966	**0.8000**	**0.8271**

3.4 Most Discriminative Regions

Different parts of the brain are responsible for different tasks and not all regions are related to the early AD. Therefore, we analyze the brain functional connectivity of different stages of early AD to search for these relevant ROIs for a better understanding of brain abnormalities. The dynamic FCN has been used in this paper, and then we can utilize the significance test [23] of functional connectivity to calculate the top 10 brain region connectivity that most related to the corresponding classification task. Figure 4 shows the top 10 connected ROIs. Figure 5 shows the brain regions corresponding to top 10 connected ROIs. Table 4 lists the top 10 most frequent ROIs for our six classification tasks. The results show that inferior temporal gyrus (ITG.R) [24], middle frontal gyrus (MFG.L) [25], hippocampus (HIP.L) [26] and Inferior frontal gyrus (IFGtriang.L) [27] are highly related to the detection of early AD.

Table 4. Top 10 most discriminative ROIs

No.	ROI index	ROI abbr	No.	ROI index	ROI abbr
1	90	ITG.R	6	58	PoCG.R
2	7	MFG.L	7	68	PCUN.R
3	37	HIP.L	8	75	PAL.L
4	13	IFGtriang.L	9	56	FFG.R
5	78	THA.R	10	20	SMA.R

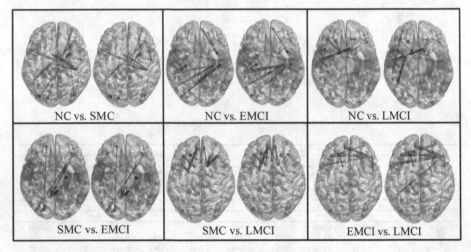

Fig. 4. Top 10 brain region connectivity.

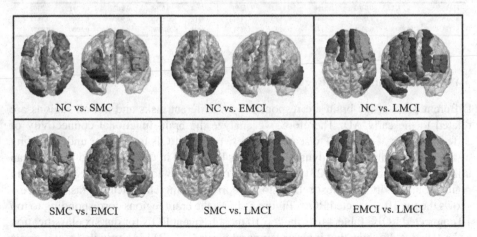

Fig. 5. Brain region corresponding to the top 10 connected ROIs.

4 Conclusion

In this paper, we propose a SFG U-Net to detect the early AD. The SFG U-Net can extract the high-order and low-order information of the graph, and the pool layer in the model can consider both graph structure and node feature to select the informative nodes for constructing a new graph. The experimental results demonstrate that our method is better than other popular methods, which indicates that our model could be meaningful to assist clinicians detecting early AD.

Acknowledgement. This work was supported partly by National Natural Science Foundation of China (Nos. 6210010638), China Postdoctoral Science Foundation (Nos. 2019M653014).

References

1. Alzheimer's Association: 2019 Alzheimer's disease facts and figures. Alzheimer's Dement. **15**, 321–387 (2019)
2. Risacher, S.L., Kim, S., Nho, K., Foroud, T., Shen, L., Petersen, R.C., et al.: APOE effect on Alzheimer's disease biomarkers in older adults with significant memory concern. Alzheimer's Dement. **11**, 1417–1429 (2015)
3. Gauthier, S., Reisberg, B., Zaudig, M., Petersen, R.C., Ritchie, K., Broich, K., et al.: Mild cognitive impairment. Lancet **367**, 1262–1270 (2006)
4. Hampel, H., Lista, S.J.: The rising global tide of cognitive impairment. Nat. Rev. Neurol. **12**, 131–132 (2016)
5. Li, Y., Liu, J., Tang, Z., Lei, B.J.: Deep spatial-temporal feature fusion from adaptive dynamic functional connectivity for MCI identification. IEEE Trans. Med. Imaging **39**, 2818–2830 (2020)
6. Li, Y., Liu, J., Gao, X., Jie, B., Kim, M., Yap, P.-T., et al.: Multimodal hyper-connectivity of functional networks using functionally-weighted LASSO for MCI classification. Med. Image Anal. **52**, 80–96 (2019)
7. Lei, B., Yang, P., Wang, T., Chen, S., Ni, D.: Relational-regularized discriminative sparse learning for Alzheimer's disease diagnosis. IEEE Trans. Cybern. **47**, 1102–1113 (2017)
8. Yang, P., Zhou, F., Ni, D., Xu, Y., Lei, B.J.: Fused sparse network learning for longitudinal analysis of mild cognitive impairment. IEEE Trans. Cybern. **47**, 1–14 (2019)
9. Huettel, S.A., Song, A.W., McCarthy, G.: Functional Magnetic Resonance Imaging, vol. 1. Sinauer Associates, Sunderland (2004)
10. Chen, B., Wang, S., Sun, W., Shang, X., Liu, H., Liu, G., et al.: Functional and structural changes in gray matter of Parkinson's disease patients with mild cognitive impairment. Eur. J. Radiol. **93**, 16–23 (2017)
11. Uddin, L.Q., Clare Kelly, A., Biswal, B.B., Xavier Castellanos, F., Milham, M.P.J.: Functional connectivity of default mode network components: correlation, anticorrelation, and causality. Hum. Brain Mapp. **30**, 625–637 (2009)
12. Arbabshirani, M.R., Damaraju, E., Phlypo, R., Plis, S., Allen, E., Ma, S., et al.: Impact of autocorrelation on functional connectivity. Neuroimage **102**, 294–308 (2014)
13. Mu, Y., Liu, X., Wang, L.J.: A Pearson's correlation coefficient based decision tree and its parallel implementation. Inf. Sci. **435**, 40–58 (2018)
14. Schwab, S., Harbord, R., Zerbi, V., Elliott, L., Afyouni, S., Smith, J.Q., et al.: Directed functional connectivity using dynamic graphical models. Neuroimage **175**, 340–353 (2018)

15. Song, X., et al.: Graph convolution network with similarity awareness and adaptive calibration for disease-induced deterioration prediction. Med. Image Anal. **69**, 101947 (2021)
16. Gao, H., Ji, S.: Graph U-Nets. In: International Conference on Machine Learning, pp. 2083–2092 (2019)
17. Lostar, M., Rekik, I.J.: Deep Hypergraph U-Net for Brain Graph Embedding and Classification. arXiv preprint arXiv:13118 (2020)
18. Wang, J., Ma, A., Chang, Y., Gong, J., Jiang, Y., Qi, R., et al.: scGNN is a novel graph neural network framework for single-cell RNA-Seq analyses. Nat. Commun. **12**, 1–11 (2021)
19. Ronneberger, O., Fischer, P., Brox, T.: U-Net: convolutional networks for biomedical image segmentation. In: International Conference on Medical Image Computing and Computer-Assisted Intervention, pp. 234–241 (2015)
20. Kipf, T.N., Welling, M.J.: Semi-supervised classification with graph convolutional networks. arXiv preprint arXiv:02907 (2016)
21. Parisot, S., Ktena, S.I., Ferrante, E., Lee, M., Guerrero, R., Glocker, B., et al.: Disease prediction using graph convolutional networks: application to autism spectrum disorder and Alzheimer's disease. Med. Image Anal. **48**, 117–130 (2018)
22. Feng, Y., You, H., Zhang, Z., Ji, R., Gao, Y.: Hypergraph neural networks. In: Proceedings of the AAAI Conference on Artificial Intelligence, pp. 3558–3565 (2019)
23. Farris, J.S., Kallersjo, M., Kluge, A.G., Bult, C.J.: Constructing a significance test for incongruence. Syst. Biol. **44**, 570–572 (1995)
24. Chertkow, H., Bub, D.J.: Semantic memory loss in dementia of Alzheimer's type: what do various measures measure? Brain **113**, 397–417 (1990)
25. Zhang, Y., Simon-Vermot, L., Caballero, M.Á.A., Gesierich, B., Taylor, A.N., Duering, M., et al.: Enhanced resting-state functional connectivity between core memory-task activation peaks is associated with memory impairment in MCI. Neurobiol. Aging **45**, 43–49 (2016)
26. Zanchi, D., Giannakopoulos, P., Borgwardt, S., Rodriguez, C., Haller, S.J.: Hippocampal and amygdala gray matter loss in elderly controls with subtle cognitive decline. Front. Aging Neurosci. **9**, 50 (2017)
27. Lin, F., Ren, P., Lo, R.Y., Chapman, B.P., Jacobs, A., Baran, T.M., et al.: Insula and inferior frontal gyrus' activities protect memory performance against Alzheimer's disease pathology in old age. J. Alzheimer's Dis. **55**, 669–678 (2017)

A Method for Predicting Alzheimer's Disease Based on the Fusion of Single Nucleotide Polymorphisms and Magnetic Resonance Feature Extraction

Yafeng Li, Yiyao Liu, Tianfu Wang, and Baiying Lei[✉]

National-Region Key Technology Engineering Laboratory for Medical Ultrasound, Guangdong Key Laboratory for Biomedical Measurements and Ultrasound Imaging, School of Biomedical Engineering, Health Science Center, Shenzhen University, Shenzhen 518060, China
leiby@szu.edu.cn

Abstract. The application of imaging genomics to the detection of Alzheimer's disease and the analysis of causative factors is of relevance. However, traditional studies are usually based on imaging data, thus neglecting the disease information implied by genetic data. In this paper, based on magnetic resonance imaging (MRI) data and single nucleotide polymorphism (SNP) data, we innovatively propose a novel data feature extraction method and construct a multimodal data fusion analysis model. Firstly, we pre-screen the SNP data and use the encoding layer in the transformer model for the screened SNP data to obtain the position information of each SNP in the sequence through the position encoding module, and then capture the features of the SNPs through the multi-head attention mechanism. Next, we perform feature extraction on the pre-processed MRI data. We adopt the idea of soft thresholding to extract the most discrepant features possible. To this end, we build a feature extraction module for MRI data that combines a soft thresholding module and a CNN module. Finally, we stitch together image features and genetic features and use a fully connected layer for classification. Through feature data fusion, our model was applied to multi-task analysis to identify AD patients, predict AD-associated brain regions, and analyse out strong correlation pairs between brain regions and risk SNPs. In multimodal data experiments, our proposed model showed better classification performance and pathogenic factor prediction, providing a new perspective for the diagnosis of AD.

Keywords: Alzheimer's disease · MRI · SNP · Transformer · Soft thresholding module

1 Introduction

Alzheimer's disease (AD) is a chronic neurodegenerative disease [1, 2] whose main clinical manifestations are memory impairment, behavioral disturbances and cognitive decline [3], which severely affects the quality of life of patients. Furthermore, according to the Alzheimer's Association (2021), there are more than 56 million people with AD

© Springer Nature Switzerland AG 2021
T. Syeda-Mahmood et al. (Eds.): ML-CDS 2021, LNCS 13050, pp. 105–115, 2021.
https://doi.org/10.1007/978-3-030-89847-2_10

worldwide, and it is expected to reach 152 million by 2050 [4]. The mild cognitive impairment (MCI) stage is a stage in the progression to AD [5]. MCI is further divided into light MCI (LMCI, who suffers from light MCI) and stable MCI (sMCI, whose symptoms are stable and will not progress to AD in 18 months). Therefore, we not only want to predict the occurrence of AD, but also want to explore the genetic and brain area links between AD and NC, AD and sMCI, AD and LMCI.

With the advancement and development of imaging technology, scientists have started to apply imaging technology to the diagnosis of AD with success [6]. Structural MRI can be used to predict AD conditions as it is very sensitive to neuronal degeneration in patients [7, 8]. On the other hand, researchers have applied well-established Genome-Wide Association Studies (GWAS) techniques in the field of AD and have analyzed risk genes associated with AD [9]. With the continuous development of imaging and genomics, scholars have used imaging genomics to better explore the links between endophenotype, ectophenotype and disease [10].

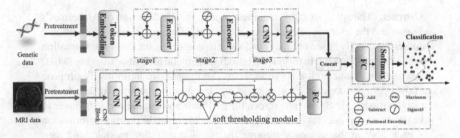

Fig. 1. The flowchart of the proposed method.

Genetic data is divided into various kinds, among which the common data used in this research field is generally Single Nucleotide Polymorphism (SNP) data. Therefore, how to select SNP data and whether significant features can be extracted from SNP data is very important for the subsequent development of the experiment. Huang *et al.* extracted SNPs from known AD risk genes and within 20 kb of the risk gene boundary as genetic data for their experiment [11]. Bi *et al.* first screened genes within the range with SNP numbers greater than a certain threshold as candidate genes, and then extracted the SNP sequences under that gene and intercepted the already extracted SNP sequences according to the set threshold range [12]. Eun Ae Kang et al. performed the association analysis of SNP first by GWAS technique to extract the significant SNP to the disease by association [13]. For disease prediction, deep learning has been successful in predicting AD [14, 15]. Basaia et al. successfully used CNN networks to classify and predict AD patients [16].

The current research has not yet found a suitable paradigm for extracting features from genetic data, and our study provides a novel way of extracting features from genetic data, i.e., using the multi-headed attention mechanism of transformer [17] to extract features from genetic data. In addition, in our experiments, we also apply the soft thresholding to extract features from MRI data. Finally, we fuse the features in the classification task, and our network achieves good classification results. The flowchart of this paper is shown in Fig. 1, and the general description of the model framework is as follows:

1) multimodal data pre-processing; 2) feature extraction and fusion; and 3) diseases classification.

2 Material and Methodology

2.1 Data Acquisition and Preprocessing

The data for this research work are obtained from the Alzheimer's Disease Neuroimaging Initiative (ADNI) database. In this study, we use 814 MRI data from the ADNI1 database, where 728 samples have both genetic data and imaging data based on the screening and matching information. The genetic data of the 728 samples were obtained from the Human 610-Quad Bead Chip in the ADNI data. Based on the pre-processing operation of the genetic data, only 691 cases are used and information on sex and age of the 691 cases is shown in Table 1. We perform separate pre-processing operations on MRI data and SNP data.

Table 1. Individual information form

Label	Number	Age (years) (Mean ± SD)	Gender (M/F)	Weight (kg) (Mean ± SD)
AD	161	75.48 ± 7.48	86/75	72.18 ± 13.65
NC	193	75.78 ± 4.92	106/87	76.40 ± 15.31
sMCI	207	75.40 ± 7.25	132/75	75.89 ± 14.56
LMCI	130	74.64 ± 6.77	89/41	76.85 ± 13.78

For magnetic resonance imaging data, we use centroid method to relocate all MRI image data, and use statistical parameter mapping (SPM8) tool to perform preprocessing. The specific steps are as follows: 1) Correct the head movement and geometric distortion. 2) Use the graph-cut method to conduct skull stripping. 3) Registration of MRI images using the International Brain Mapping Consortium template. 4) Segment the corresponding anatomical regions into gray matter (GM), white matter (WM), and cerebrospinal fluid (CSF). 5) Resample these images to a quadratic resolution of 1.5 mm. 6) The surface of these images was smoothed using a 60 mm full-width in a half-maximum Gaussian kernel. 7) Use a toolbox for data processing and analysis for brain imaging to obtain the GM tissue. 8) 116 regions of interest (ROI) were obtained by automated anatomical labeling (AAL), and 26 cerebellar tissues were removed. The final ROIs of 90 brain regions were retained.

SNP data pre-processing operations. We used plink software for quality control of the genetic data. The steps are as follows: 1) Filter SNPs and individuals based on a threshold. 2) Check sex discrepancy. 3) Minor allele frequency (MAF). 4) Hardy–Weinberg equilibrium (HWE). 5) Excludes individuals with high or low heterozygosity rates. 6) Delete relative sample. 7) Population stratification.

The final SNP data information for 691 subjects was retained following quality control procedures.

2.2 Feature Extraction

Feature Extraction of SNP Data. First, we obtain the known AD risk genes from the AlzGene database (http://www.alzgene.org/), 47 in total (see accompanying information table). Within 20kb of the risk gene boundaries, we extract SNP information, obtaining a total of 742 SNP. In stage1 and stage2, the encoding layer from the transformer, we used the position encoding module to obtain the position information of each SNP in the sequence, and then noticed the context information of the SNP sequence through the multi-headed attention mechanism to better extract the SNP sequence features. However, since the number of features extracted from the coding layer was too large, we added a stage3 convolutional layer to further filter the features. stage3 contains a convolutional layer, a batch layer, a poolayer and a relu activation function. Figure 2 shows a diagram of our genetic data feature extraction module.

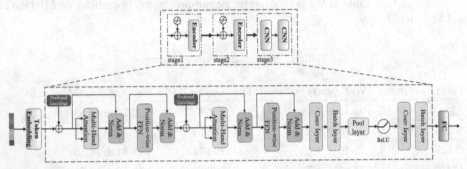

Fig. 2. Schematic diagram of the gene data feature extraction module. We use two encoders, each contains multiple attentions. Finally, we use two convolution modules to extract further features.

Feature Extraction of MRI Data. When extracting features from image data, we reduce the noise in the data and extract as many features as possible. Soft thresholding is a classical method used in the field of signal noise reduction, which effectively filters out some noise from the signal [18] and removes redundant features. Soft thresholding shrinks the input data towards zero with the following equation.

$$y = \begin{cases} x - \tau & x > \tau \\ 0 & -\tau \leq x \leq \tau \\ x + \tau & x < -\tau \end{cases} \tag{1}$$

where x denotes the input MRI data, y denotes the output extracted MRI features and τ denotes the threshold value.

Our feature extraction from image data module allows the model to learn the difference between the shallow and deep information. It takes into account the shallow and deep information extracted by the convolutional layer, and assigns this difference information to the original information, thus allowing the model to obtain the internal difference information of the features. We first perform a preliminary feature extraction of the input values X_{input} using the CNN module. The features f_1 extracted by the CNN_1 module are used as the base values for the thresholds. Let the features f_2 extracted by the CNN_3 module obtained by the sigmoid activation function, after which they will be f_3 multiplied with f_1, f_4 as the subtracted value in the next step. Subtract the value f_4 from the base value f_1 and then let the results be subtracted from each other to obtain the zero-value matrix f_5. The subtracted value is compared with the zero value to obtain the larger of the two f_6 The larger value is multiplied with the activated feature to obtain feature f_7, and finally f_7 is summed with the input value X_{input} to obtain the output value X_{output}. Figure 3 shows a schematic of our image data feature extraction module.

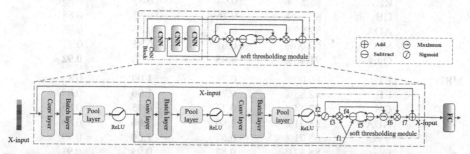

Fig. 3. This is a schematic of the image data feature extraction module. There are three convolution modules, each contains a convolution layer, a batch normalization layer, a pooling layer, and a ReLU activation function. On the right is our soft threshold residual shrinkage module.

Fusion of Feature Data and Classification. We performed a stitching operation on features extracted from genetic data and features extracted from imaging data to obtain our fused features. This was then passed through a fully-connected layer and a Softmax classifier to implement our disease prediction classification.

3 Experimental Results

3.1 Model Performance and Method Comparison

In the experiments, our model are effective for AD and NC classification, with an accuracy of 91.43%, while the prediction accuracy with SNP data and MRI data was 68.57% and 88.57%, respectively. We put the SNP data and MRI data into machine learning models (Extra Trees, ET; Random Forest, RF; AdaBoost parameters, ADA; Gradient Boosting, GB; Support Vector Classifier. SVC) for five-fold cross-validation training. The experimental results show that our method has some improvement in classifica-tion performance over traditional machine learning methods. Table 2 contains the classification performance of our model and the machine learning model under the AD, NC classification task with fused data and unimodal data. Table 3 contains the classification performance of our model and the machine learning model for the AD, sMCI classification task with both fused and unimodal data. Table 4 contains the classification performance of our model and the machine learning model for the AD, LMCI classification task with both fused and unimodal data.

Figures 4, 5, and 6 respectively show the classification performance of our model and other machine learning models for fusion data and monomodal data under three classification tasks.

Table 2. AD VS NC classification performance.

Data	Model	Accuracy	Precision	Recall	F1-score
Fusion	ET	71.80%	0.67	**1.00**	0.80
	RF	88.57%	0.94	0.85	0.89
	GB	80.00%	0.76	0.95	0.84
	ADA	82.85%	0.79	0.95	0.86
	SVC	60.00%	0.61	0.85	0.71
	Ours	**91.43%**	**0.95**	0.90	**0.92**
MRI	ET	80.00%	0.74	**1.00**	0.85
	RF	82.85%	0.77	**1.00**	0.87
	GB	77.14%	0.73	0.95	0.83
	ADA	85.71%	0.80	**1.00**	0.89
	SVC	65.71%	0.62	**1.00**	0.77
	Ours	**88.57%**	**0.87**	0.95	**0.91**
SNP	ET	60.00%	0.59	**1.00**	**0.74**
	RF	60.00%	0.59	**1.00**	**0.74**
	GB	54.28%	0.56	0.95	0.70
	ADA	54.28%	0.56	0.90	0.69
	SVC	60.00%	0.61	0.85	0.71
	Ours	**68.57%**	**0.80**	0.60	0.69

3.2 Abnormal Brain Regions and Pathogenic Genes

By analyzing the image features extracted by the model under different classification tasks, we obtained brain regions highly associated with AD, such as the Hippocampus, Superior temporal gyrus and other brain regions that have been shown to be associated with AD. For example, in the AD, NC classification task, the left side of the Posterior cingulate gyrus ($p < 0.01$), the left side of the Hippocampus ($p < 0.01$), the right side of the Hippocampus ($p < 0.01$), the left side of the Superior temporal gyrus ($p < 0.01$), etc., while in the sMCI and LMCI categorization tasks, most brain region abnormalities were not significantly represented, except for Lenticular nucleus, putamen ($p < 0.05$, AD vs sMCI), and Posterior cingulate gyrus ($p < 0.05$, AD vs LMCI). We put the results of the top 8 relatively abnormal brain regions for each type of task on display in Fig. 7. We have placed the structural connectivity maps of brain regions for the three categorization tasks in Fig. 8.

We also analysed the characteristics of the SNP data extracted by the model. In the AD, NC classification task, SNPs such as rs1323720 ($p < 0.01$), rs1614735 ($p < 0.01$), rs1133174 ($p < 0.01$) had a greater effect on the onset of AD in individuals, and in the sMCI, AD classification task, SNPs from the DAPK1 gene on rs2418828 ($p < 0.01$), rs7855635 ($p < 0.01$), rs2058882 ($p < 0.01$), and rs10125534 ($p < 0.01$) were more

Table 3. AD VS sMCI classification performance.

Data	Model	Accuracy	Precision	Recall	F1-score
Fusion	ET	58.33%	0.58	**1.00**	0.74
	RF	61.11%	0.61	0.95	0.74
	GB	52.78%	0.56	0.86	0.68
	ADA	50.00%	0.55	0.86	0.68
	SVC	61.12%	0.60	**1.00**	**0.75**
	Ours	**63.89%**	**0.63**	0.90	**0.75**
MRI	ET	58.33%	0.59	0.90	0.72
	RF	61.11%	0.62	0.86	0.72
	GB	52.78%	0.57	0.81	0.67
	ADA	50.00%	0.55	0.81	0.65
	SVC	58.33%	0.58	**1.00**	0.74
	Ours	**69.44%**	**0.69**	0.86	**0.77**
SNP	ET	**61.11%**	0.60	**1.00**	**0.75**
	RF	58.33%	0.58	**1.00**	0.74
	GB	**61.11%**	0.60	**1.00**	**0.75**
	ADA	58.33%	0.58	**1.00**	**0.75**
	SVC	**61.11%**	0.60	**1.00**	**0.75**
	Ours	58.33%	**0.64**	0.67	0.65

Table 4. AD VS LMCI classification performance.

Data	Model	Accuracy	Precision	Recall	F1-score
Fusion	ET	55.17%	0.43	0.55	0.48
	RF	62.06%	0.50	0.55	0.52
	GB	65.51%	0.53	**0.91**	**0.67**
	ADA	55.17%	0.45	0.82	0.58
	SVC	48.27%	0.41	0.82	0.55
	Ours	**68.97%**	**0.63**	0.45	0.53
MRI	ET	51.72%	0.40	0.55	0.46
	RF	44.82%	0.39	0.82	0.53
	GB	48.27%	0.42	**0.91**	0.57
	ADA	51.72%	0.43	0.82	0.56
	SVC	62.06%	0.00	0.00	0.00
	Ours	**72.41%**	**0.62**	0.73	**0.67**
SNP	ET	58.62%	0.46	0.55	0.50
	RF	58.62%	0.44	0.36	0.40
	GB	37.93%	0.33	0.64	0.44
	ADA	44.82%	0.39	**0.82**	0.53
	SVC	51.72%	0.43	**0.82**	**0.56**
	Ours	**68.97%**	**0.63**	0.45	0.53

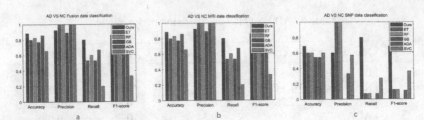

Fig. 4. Experimental performance plot of our model against other machine learning models under the AD VS NC classification task.

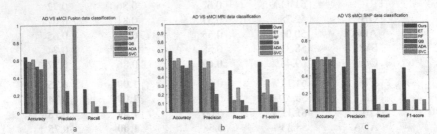

Fig. 5. Experimental performance plot of our model against other machine learning models under the AD VS sMCI classification task.

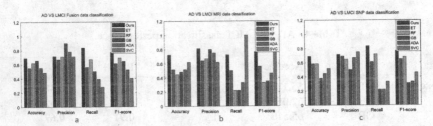

Fig. 6. Experimental performance plot of our model against other machine learning models under the AD VS LMCI classification task.

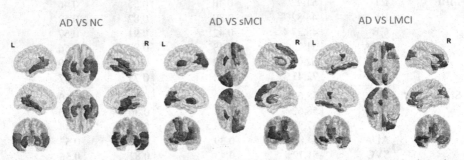

Fig. 7. Map of abnormal brain areas. From left to right, the distribution of abnormal brain regions in AD vs NC, AD vs sMCI, AD vs LMCI, and the three tasks is shown.

potent for disease in the sMCI, AD classification task, while rs2276346 (p < 0.01), located on the SORL1 gene, had a greater effect on LMCI. Figure 9 shows a circle plot of our 8 abnormal brain regions and 8 risk SNP.

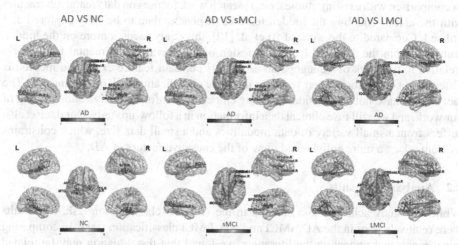

Fig. 8. Structural connectivity map of abnormal brain regions.

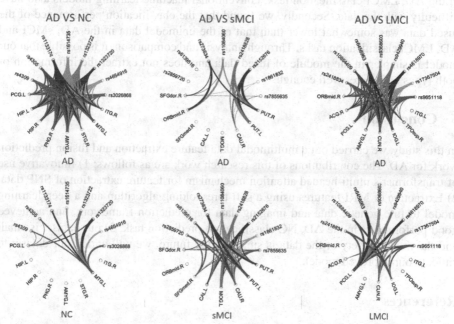

Fig. 9. Top 8 abnormal brain regions and top 8 risk SNP connectivity circles for each type of task.

4 Discussion

4.1 Comparison with Existing Studies

In comparison with existing studies, our research work focuses on data feature extraction with the aim of enabling the hidden features of genetic data to be fully explored and utilised. Compared to the work of Bi et al. [12], this study focuses more on the hidden features within the data and lacks the design of the overall experimental framework. Relative to the work of Venugopalan et al. [14] our data feature extraction method is different but outperforms their work for the same classification task of AD and NC. The fact that we have not yet considered fusing clinical data information is a shortcoming of our work, and we will fuse clinical data information in a follow-up study. Our dataset also suffers from a small variety of data modalities and a small data size, which constrains us from taking a more enlightened view of the causative factors of AD.

4.2 Analysis of Results

While our study achieved good results in the AD, NC classification task, the results were relatively poor in the AD, sMCI and AD, LMCI classification tasks. In comparing experiments and reviewing the literature, we found that this situation may be related to the type and quality of the data itself. Both in the AD, sMCI classification task and in the AD, LMCI classification task, conventional machine learning models also have difficulty with the data. Secondly, we found that the classification performance of the fused data was somewhat lower than that of the unimodal data in the AD, sMCI and AD, LMCI classification tasks. Through analysis and comparison, it is possible that our model is simpler in the module of fused data and does not extract the information of both types of features well enough.

5 Conclusion

In this study, we carried out a multimodal data feature extraction and fusion prediction work for AD. The contributions of this research work are as follows: 1) Innovative use of transformer's multi-headed attention mechanism for feature extraction of SNP data. 2) Extraction of MRI features using a soft thresholding algorithm and a deep learning model to fuse genetic data and imaging data for prediction framework, and achieved good performance in the AD, NC classification prediction task. 3) our dataset is small and the modal variety of the data is small. In the future, we will enrich our dataset to validate our improved model.

References

1. Khachaturian, Z.S.: Diagnosis of Alzheimer's disease. J. Arch. Neurol. **4211**, 1097–1105 (1985)
2. Kucmanski, L.S., Zenevicz, L., Geremia, D.S., Madureira, V.S.F., Silva, T.G.D., Souza, S.S.D.: Alzheimer's disease: challenges faced by family caregivers. J. Revista Brasileira de Geriatria e Gerontologia **196**, 1022–1029 (2016)

3. Mucke, L.: Alzheimer's disease. Nature **4617266**, 895–897 (2009)
4. Alzheimer's Association: 2021 Alzheimer's disease facts and figures. J. Alzheimer's Dement. **173**, 327–406 (2021)
5. Chen, X., Zhang, H., Gao, Y., Wee, C.Y., Li, G., Shen, D., et al.: High-order resting-state functional connectivity network for MCI classification. J. Hum. Brain Mapp. **379**, 3282–3296 (2016)
6. Marchitelli, R., Aiello, M., Cachia, A., Quarantelli, M., Cavaliere, C., Postiglione, A., et al.: Simultaneous resting-state FDG-PET/fMRI in Alzheimer disease: relationship between glucose metabolism and intrinsic activity. J. Neuroimage **176**, 246–258 (2018)
7. McEvoy, L.K., Fennema-Notestine, C., Roddey, J.C., Hagler, D.J., Jr., Holland, D., Karow, D.S., et al.: Alzheimer disease: quantitative structural neuroimaging for detection and prediction of clinical and structural changes in mild cognitive impairment. J. Radiol. **2511**, 195–205 (2009)
8. Vemuri, P., Jack, C.R.: Role of structural MRI in Alzheimer's disease. J. Alzheimer's Res. Ther. **24**, 1–10 (2010)
9. Feulner, T., Laws, S., Friedrich, P., Wagenpfeil, S., Wurst, S., Riehle, C., et al.: Examination of the current top candidate genes for AD in a genome-wide association study. J. Mol. Psychiatry **157**, 756–766 (2010)
10. Hariri, A.R., Drabant, E.M., Weinberger, D.R.: Imaging genetics: perspectives from studies of genetically driven variation in serotonin function and corticolimbic affective processing. J. Biol. Psychiatry **5910**, 888–897 (2006)
11. Huang, M., Chen, X., Yu, Y., Lai, H., Feng, Q.: Imaging genetics study based on a temporal group sparse regression and additive model for biomarker detection of Alzheimer's disease. IEEE Trans. Med. Imaging **40**, 1461–1473 (2021)
12. Bi, X.-A., Hu, X., Wu, H., Wang, Y.: Multimodal data analysis of Alzheimer's disease based on clustering evolutionary random forest. IEEE J. Biomed. Health Inform. **2410**, 2973–2983 (2020)
13. Kang, E., Jang, J., Choi, C.H., Kang, S.B., Bang, K.B., Kim, T.O., et al.: Development of a clinical and genetic prediction model for early intestinal resection in patients with Crohn's disease: results from the IMPACT study. J. Clin. Med. **104**, 633 (2021)
14. Venugopalan, J., Tong, L., Hassanzadeh, H.R., Wang, M.D.: Multimodal deep learning models for early detection of Alzheimer's disease stage. J. Sci. Rep. **111**, 1–13 (2021)
15. Nguyen, M., He, T., An, L., Alexander, D.C., Feng, J., Yeo, B.T., et al.: Predicting Alzheimer's disease progression using deep recurrent neural networks. J. NeuroImage **222**, 117203 (2020)
16. Basaia, S., Agosta, F., Wagner, L., Canu, E., Magnani, G., Santangelo, R., et al.: Automated classification of Alzheimer's disease and mild cognitive impairment using a single MRI and deep neural networks. J. NeuroImage: Clin. **21**, 101645 (2019)
17. Vaswani, A., Shazeer, N., Parmar, N., Uszkoreit, J., Jones, L., Gomez, A.N., et al.: Attention is all you need. In: Advances in Neural Information Processing Systems, pp. 5998–6008 (2017)
18. Donoho, D.L.: De-noising by soft-thresholding. J. IEEE Trans. Inf. Theory **413**, 613–627 (1995)

Author Index

Printed in the United States
by Baker & Taylor Publisher Services